WHEN TIME BENDS

A Journey Through Einstein's Universe of Gravity, Speed, and the Elastic Nature of Time

Douglas Donovan

IBSN: 979-8-3495-7436-8

PART I:
SETTING THE STAGE

Chapter 1: The Clockmaker's Dilemma

In the summer of 1784, the master clockmaker John Harrison stood in his workshop in London, surrounded by the finest timepieces of his era. His marine chronometers had revolutionized navigation, solving the longitude problem that had plagued sailors for centuries. Each tick of his masterful creations marked the passage of time with unprecedented precision—or so he believed. Harrison was confident that time itself was as regular and predictable as the brass gears within his clocks. He could never have imagined that centuries later, physicists would discover something extraordinary: his clocks, no matter how precisely crafted, could not measure time itself, but only their own distorted perception of it.

This is the clockmaker's dilemma - the profound realization that time, the most fundamental aspect of our existence, is far stranger and more malleable than our daily experience suggests. It's a puzzle that would eventually overturn centuries of scientific understanding and reveal that the universe operates according to principles so counterintuitive that they seem like science fiction.

The Foundation of Certainty

To understand how revolutionary Einstein's discoveries truly were, we must first appreciate the world of physics that existed before him. For more than two centuries, the scientific community had built its understanding of the universe on what seemed like an unshakeable foundation: Isaac Newton's conception of absolute space and absolute time.

In 1687, Newton published his monumental work, *Philosophiae Naturalis Principia Mathematica* (Mathematical Principles of Natural Philosophy), which would dominate scientific thinking well into the twentieth century. Within its pages, Newton articulated a view of reality that felt intuitively correct to anyone who had ever lived on Earth. He proposed that "wherever we are in the universe, all time flows unchangingly at the same pace, in the same increments, from the infinite past into the infinite future," and that "space is uniform, and extends on into infinity".

Newton's concept was elegantly simple: time was like a great cosmic river, flowing at a constant rate throughout the entire universe. Whether you were sitting in a London coffee house, sailing across the Atlantic, or standing on the surface of the moon, time would pass at exactly the same rate. A second here was identical to a second there. An hour was an hour, regardless of location, motion, or circumstance.

This wasn't merely a philosophical position—it was a practical necessity. Newton's physics required a universal framework against which to measure motion and change. He distinguished between what he called "absolute time" and "relative time". Absolute time was the real thing—the actual flow of time itself, independent of any physical processes or measurements. Relative time, on the other hand, was how we actually measured time using clocks, sundials, or the motion of celestial bodies.

As Newton himself explained, relative times "are not the actual quantities whose names they bear but are those sensible measures of them (whether true or erroneous) that are commonly used instead of the quantities being measured". In other words, our clocks weren't measuring time itself—they were giving us approximations, like shadows cast by the true reality of absolute time.

The Clockwork Universe

This Newtonian worldview painted a picture of what philosophers called a "clockwork universe." Just as Harrison's marine chronometers operated with mechanical precision, the entire cosmos was seen as a vast, predictable machine. Planets orbited the sun in perfect ellipses, following mathematical laws that could be calculated with exquisite accuracy. Comets returned precisely when predicted. The moons of Jupiter danced in their celestial ballet exactly as Newton's equations foretold.

Within this framework, time served as the cosmic metronome, keeping perfect beat for the universal symphony of motion. It was the backdrop against which all events unfolded, as unchanging and reliable as mathematics itself. This conception was so compelling that it seemed to be confirmed by every observation and every successful prediction of Newtonian physics.

The idea of absolute time also solved what appeared to be fundamental problems in understanding motion. Newton defined the "true motion" of any object as its motion through absolute space, measured against the steady flow of absolute time. This provided a clear, unambiguous way to describe the physical world. A cannonball fired from a ship had its own absolute velocity through space, independent of the ship's motion. The Earth had its absolute motion around the sun, which had its own absolute motion through the galaxy.

Seeds of Doubt

Yet even in Newton's era, not everyone was entirely convinced. Throughout history, alternative views had suggested that space and time might not be fundamental entities but rather relationships between objects and events. These "relationalist" perspectives argued that space was nothing more than a way of describing the arrangements of matter, and time was simply a measure of change and motion.

The German philosopher Gottfried Leibniz, Newton's contemporary and rival, championed this relationalist view. In his famous correspondence with Samuel Clarke (who defended Newton's position), Leibniz argued that absolute space and time were unnecessary metaphysical baggage. For Leibniz, time without change was meaningless, and space without objects was empty abstraction.

But Newton had powerful arguments on his side. His famous bucket experiment seemed to demonstrate the reality of absolute space: when a bucket of water is spun, the water's surface becomes curved due to centrifugal force, even when the water is rotating at the same rate as the bucket. This suggested that there must be some absolute standard of rotation—absolute space—against which true motion could be measured.

Similarly, Newton's physics achieved such remarkable success in predicting everything from planetary orbits to tidal forces that questioning his fundamental assumptions seemed not just unnecessary but foolish. If absolute time and space weren't real, how could his equations work so perfectly?

Yet as the centuries passed, subtle cracks began to appear in this seemingly perfect edifice.

The Precision Problem

By the nineteenth century, clockmakers had achieved extraordinary precision. Harrison's chronometers, remarkable as they were, seemed crude compared to the astronomical observations and laboratory measurements that nineteenth-century scientists could perform. Astronomers could predict eclipses to the second, and physicists could measure the properties of light and matter with unprecedented accuracy.

But this very precision began to reveal puzzling discrepancies. The orbit of Mercury, for instance, didn't behave exactly as Newton's equations predicted. The difference was tiny—just 43 arcseconds per century—but it was real and measurable. If Newton's physics were perfect and time truly flowed absolutely, why did such anomalies exist?

Even more puzzling were the results of experiments involving light. By the late 1800s, physicists had established that light was a wave phenomenon, similar to sound or water waves. But waves require a medium through which to travel. Sound waves travel through air, water waves through water—so what did light waves travel through?

The answer seemed to be a mysterious substance called the "luminiferous ether"—an invisible medium that was thought to permeate all of space. If this ether existed, then it would provide an absolute reference frame for measuring motion. Objects moving through the ether should experience an "ether wind," just as a moving car experiences air resistance.

In the 1880s, Albert Michelson and Edward Morley designed an ingenious experiment to detect this ether wind by measuring the speed of light in different directions. According to

Newton's physics and the prevailing understanding of waves, light should travel faster when moving with the ether wind and slower when moving against it.

The results were shocking: no ether wind was detected. No matter which direction they measured, light always traveled at exactly the same speed. This seemed impossible according to Newtonian physics, where velocities should simply add together.

The Deepening Mystery

As the nineteenth century drew to a close, other strange phenomena began to accumulate. Experiments with fast-moving particles revealed behaviors that didn't quite fit Newtonian expectations. Observations of stellar aberration—the apparent shifting of star positions due to Earth's motion—yielded results that were difficult to reconcile with absolute space and time.

Most mysteriously of all, careful measurements began to suggest that the relationship between energy and mass might not be as straightforward as Newton had assumed. When objects moved at very high speeds—speeds approaching the velocity of light—their behavior became increasingly bizarre.

These weren't dramatic departures from Newtonian physics. They were subtle, often requiring the most sensitive instruments available to detect. But they were consistent, reproducible, and deeply troubling to physicists who had built their careers on the seemingly solid foundation of absolute space and time.

The clockmaker's dilemma was becoming apparent: no matter how precisely we built our instruments, no matter how carefully we measured time and space, something fundamental seemed to be eluding us. Our most accurate clocks and our most sophisticated experiments were revealing that the universe might not operate according to the simple, absolute principles that had seemed so obviously correct.

The Stage is Set

By the dawn of the twentieth century, physics found itself in a curious position. On one hand, Newtonian mechanics continued to achieve remarkable success in countless applications, from engineering to astronomy. On the other hand, a growing collection of anomalies suggested that something was fundamentally wrong with our understanding of space and time.

It was into this world of subtle contradictions and deepening mysteries that a young patent clerk named Albert Einstein would soon step. Armed with thought experiments that would make John Harrison's precision instruments seem almost primitive by comparison, Einstein would discover that the clockmaker's dilemma had a solution—but one so radical that it would forever change our understanding of reality itself.

The clocks in Harrison's workshop measured time faithfully, but they were measuring something far stranger than he could have imagined. Time wasn't the steady, absolute flow that Newton had envisioned. Instead, it was elastic, stretchable, and intimately connected to space in ways that would soon revolutionize physics.

The stage was set for a revolution that would reveal time dilation—the extraordinary phenomenon that would show us that time itself could slow down, speed up, and bend in response to motion and gravity. But first, we must understand how one young physicist managed to see through centuries of seemingly obvious assumptions to discover the true nature of time.

The clockmaker's dilemma was about to be resolved, but the solution would be far more beautiful and stranger than anyone had dared to imagine. In questioning the absolute nature of time, Einstein would discover something even more profound: that time and space are woven together into a single, dynamic fabric that shapes the very structure of our universe.

As we prepare to follow Einstein's revolutionary journey, we carry with us the same fundamental question that puzzled the finest minds of the nineteenth century: If our most precise measurements can't reveal absolute time, what exactly are our clocks measuring? The answer, as we shall see, lies not in building better clocks, but in understanding the true nature of time itself—a nature far more wonderful and mysterious than the steady tick of even John Harrison's finest chronometers.

Chapter 2: Einstein's Revolutionary Insight

In the spring of 1896, a sixteen-year-old student named Albert Einstein was walking through the Swiss countryside near Aarau, lost in thought about a peculiar question that would haunt him for nearly a decade. What would it be like, he wondered, to ride alongside a beam of light, traveling at the same speed as light itself? According to the physics of his day, if you could somehow match the speed of light, the electromagnetic wave should appear frozen in space—like a stationary pattern of electric and magnetic fields suspended in the ether.

But this seemed impossible. Einstein knew that such a frozen light wave violated everything that physicists understood about electromagnetic radiation. James Clerk Maxwell's elegant equations, which had successfully described all known electrical and magnetic phenomena, simply didn't allow for stationary electromagnetic fields of this type. The mathematics broke down completely.

This teenage thought experiment would prove to be the first crack in the foundation of classical physics—a crack that would eventually split open to reveal an entirely new understanding of space, time, and reality itself. What Einstein couldn't have known as he walked those Swiss paths was that his innocent question about riding a light beam would launch him on a journey that would revolutionize human understanding of the universe.

The Long Road to Discovery

Einstein's path to relativity was neither straight nor swift. The discovery was not momentary but rather, in Einstein's own reckoning, the outcome of seven and more years of intense intellectual struggle. From that initial thought experiment in 1896 to the publication of special relativity in 1905, Einstein grappled with contradictions that seemed to strike at the heart of physics itself.

The young Einstein was particularly troubled by what appeared to be a fundamental conflict between two pillars of nineteenth-century physics. On one side stood the principle of relativity, inherited from Galileo and codified by Newton, which stated that the laws of physics should be the same in all uniformly moving reference frames. On the other side stood Maxwell's electromagnetic theory, which seemed to predict that light always traveled at a specific speed—but a speed relative to what?

According to the prevailing wisdom, light waves traveled through the luminiferous ether at a fixed speed relative to that mysterious medium. But this created a disturbing asymmetry in the laws of physics. Mechanical phenomena obeyed the principle of relativity—you could perform mechanical experiments inside a smoothly moving train and get the same results as if the train were stationary. But electromagnetic phenomena seemed to provide a way to detect "absolute motion" through the ether.

This bothered Einstein deeply. As he later reflected, "the asymmetries which do not appear to be inherent in the phenomena" suggested that something was fundamentally wrong with the accepted picture. Why should the universe treat mechanical and electromagnetic phenomena so differently?

The Michelson-Morley Puzzle

While Einstein wrestled with these theoretical concerns, experimental physics was uncovering results that seemed to defy explanation. The Michelson-Morley experiment of 1887 had attempted to detect Earth's motion through the ether by measuring differences in the speed of light in different directions. The experiment was exquisitely designed and precisely executed—and it found absolutely nothing.

No matter how the apparatus was oriented, no matter what time of year the measurements were taken, light always traveled at exactly the same speed in all directions. This was deeply puzzling because Earth's orbital motion should have created detectable variations in the speed of light if the ether existed as physicists imagined it.

Various explanations were proposed to save the ether theory. Perhaps the Earth dragged the ether along with it, or perhaps moving objects contracted in just the right way to hide the ether wind. The Dutch physicist Hendrik Lorentz and the Irish physicist George FitzGerald independently suggested that objects might physically shrink in the direction of motion through the ether by exactly the amount needed to explain the null result of the Michelson-Morley experiment.

But these ad hoc explanations troubled Einstein. They seemed contrived, designed more to preserve existing theory than to reveal fundamental truth. There had to be a deeper, more elegant explanation—one that would resolve the apparent conflict between the principle of relativity and electromagnetic theory.

The Breakthrough Insight

The key breakthrough came when Einstein realized that the problem lay not in the physics of light or motion, but in something far more fundamental: our assumptions about the nature of time itself. As he later recalled, "The difficulty to be overcome lay in the constancy of the velocity of light in a vacuum, which I first believed had to be given up. Only after years of groping did I notice that the difficulty lay in the arbitrariness of basic kinematical concepts".

The crucial insight was the relativity of simultaneity. Events that appear to occur at the same time for one observer might not be simultaneous for another observer moving at a different velocity. This wasn't just a technical detail—it struck at the heart of Newton's conception of absolute time.

Einstein realized that if he abandoned the assumption of absolute simultaneity—if he accepted that "at the same time" was not a universal concept but depended on the observer's motion—then the principle of relativity and the constancy of light speed could coexist beautifully. The price was revolutionary: space and time behaved very differently than Newton had told us.

The Miraculous Year

In 1905, working as a patent clerk in Bern, Einstein published a series of papers that would forever change physics. His paper on special relativity, titled "On the Electrodynamics of Moving Bodies," appeared almost modest in its scope. It contained no references to previous work, no dramatic claims about overturning established physics, and no acknowledgment that it was revolutionizing our understanding of space and time.

Yet within its pages lay principles that would transform human understanding of reality. Einstein built his theory on just two fundamental postulates: the principle of relativity (the laws of physics are the same in all uniformly moving reference frames) and the constancy of the speed of light (light travels at the same speed in all directions for all uniformly moving observers).

From these simple starting points, Einstein derived consequences that seemed to defy common sense. Moving clocks run slow. Moving objects contract in length. Mass and energy are interchangeable. Simultaneity is relative. The universe has a speed limit, and nothing with mass can reach it.

Perhaps most remarkably, Einstein showed that Maxwell's electromagnetic equations didn't need to be modified at all to be compatible with the principle of relativity. The apparent asymmetries that had troubled physicists were artifacts of incorrect assumptions about space and time, not real features of electromagnetic phenomena.

The Happiest Thought

But Einstein wasn't finished. In 1907, while still working at the patent office, he had what he would later call his "happiest thought". He realized that the principle of relativity could be extended beyond uniform motion to include acceleration and gravity.

The key insight was the equivalence principle: there is no way to distinguish between the effects of gravity and the effects of acceleration. A person in a windowless elevator cannot tell whether they are being pulled downward by gravity or pushed upward by the elevator's acceleration. This seemingly simple observation would lead to consequences even more profound than those of special relativity.

If gravity and acceleration are equivalent, Einstein reasoned, then gravity must affect the passage of time just as acceleration does in special relativity. In his 1907 paper expanding on this idea, Einstein predicted the phenomenon of gravitational time dilation—clocks in stronger gravitational fields should run slower than clocks in weaker fields.

This was not just a minor adjustment to existing theory. It suggested that gravity itself was not a force in the traditional sense but rather a manifestation of the geometry of spacetime. Mass and energy don't just move through space and time—they actively warp the fabric of spacetime itself.

The Mathematical Challenge

Developing these insights into a complete theory required mathematics that Einstein didn't initially possess. When he returned to Switzerland in 1912 to accept a professorship at ETH Zurich, he immediately sought out his old classmate Marcel Grossmann, now a professor of mathematics.

Grossmann introduced Einstein to the mathematical tools he needed: Riemannian geometry and the broader field of differential geometry. These mathematical frameworks, developed by mathematicians like Bernhard Riemann and Gregorio Ricci-Curbastro, provided ways to describe curved spaces and surfaces—exactly what Einstein needed to describe curved spacetime.

The collaboration between Einstein and Grossmann marked a turning point in theoretical physics. For the first time, the most advanced mathematics of the day was being directly applied to understand the fundamental structure of physical reality. The universe, it seemed, was written in a mathematical language far more sophisticated than Newton had imagined.

The Field Equations

By 1915, after years of struggle with the mathematical formalism, Einstein was ready to present his complete theory of general relativity to the world. The culmination was a set of field equations that described how matter and energy curve spacetime, and how that curvature guides the motion of matter and energy.

Einstein's journey to the final form of these equations was not without false starts. In October 1915, he published field equations that seemed promising—they correctly predicted the previously unexplained advance of Mercury's perihelion. But Einstein soon realized these equations had a fatal flaw: they violated the local conservation of energy and momentum.

Returning to the drawing board, Einstein corrected his mistake and on November 25, 1915, presented the final form of his field equations to the Prussian Academy of Sciences. These equations, elegant in their mathematical beauty, captured a revolutionary new picture of gravity: mass and energy tell spacetime how to curve, and curved spacetime tells matter how to move.

Vindication and Fame

For four years after 1915, Einstein's theory remained largely a theoretical curiosity, admired by a small circle of physicists but lacking dramatic experimental confirmation. That changed dramatically on May 29, 1919, during a total solar eclipse.

Arthur Eddington, the British astronomer who led one of the eclipse expeditions, had designed an experiment to test one of Einstein's most striking predictions: that light from distant stars should be deflected when it passes near the sun. According to Newton's theory, light should be deflected by half the amount that Einstein predicted.

The results were unambiguous. When Eddington's team measured the apparent positions of stars near the sun during the eclipse, they found deflections that matched Einstein's predictions almost exactly. The announcement of these results on November 6, 1919, at a joint meeting of the Royal Society and Royal Astronomical Society in London, made Einstein famous overnight.

The London Times proclaimed "Revolution in Science—New Theory of the Universe—Newtonian Ideas Overthrown," and Einstein became perhaps the first scientist to achieve genuine celebrity status. But more importantly for physics, the eclipse results provided dramatic confirmation that space and time really do behave as Einstein's theory predicted.

The New Cosmos

Einstein's revolutionary insights didn't stop with the confirmation of light deflection. In 1917, he applied his field equations to the universe as a whole, founding the field of relativistic cosmology. His calculations suggested that the universe couldn't be static—it must be either expanding or contracting.

This prediction seemed so contrary to the accepted picture of an eternal, unchanging cosmos that Einstein introduced a "cosmological constant" into his equations to force them to describe a static universe. He would later call this "the biggest blunder of my life," but his original instinct was correct. In 1927, Georges Lemaître found solutions to Einstein's original equations that described an expanding universe, theoretically deriving what would later be confirmed as Hubble's law.

The Transformation of Reality

Looking back, we can see that Einstein's insights between 1905 and 1915 represent one of the most profound transformations in human understanding of reality. The comfortable, intuitive world of absolute space and absolute time gave way to a strange, dynamic universe where space and time are woven together into a single fabric that can stretch, contract, and curve.

The implications extended far beyond academic physics. Einstein's work revealed that the universe operates according to principles that seem to violate common sense but are mathematically elegant and experimentally verifiable. Time is not universal but personal. Space is not fixed but flexible. Mass and energy are not separate but interchangeable.

These discoveries required Einstein to develop new ways of thinking about the most fundamental aspects of existence. His thought experiments—imagining himself riding alongside a light beam, falling freely in an elevator, watching clocks from different reference frames—became powerful tools for probing the nature of reality.

The Method Behind the Revolution

What made Einstein's approach so revolutionary was not just his willingness to question fundamental assumptions but his method of doing so. Rather than starting with experimental anomalies and trying to patch existing theory, Einstein began with deep theoretical principles and followed them wherever they led, even when they contradicted everyday experience.

He trusted mathematical beauty and theoretical consistency over intuition. When his equations predicted that time could run at different rates or that space could curve, Einstein didn't dismiss these predictions as absurd but instead asked what experiments might test them.

This approach—starting with fundamental principles and working out their consequences through thought experiments and mathematical analysis—became a model for theoretical physics that continues to this day. Einstein showed that the human mind, armed with mathematics and guided by physical principles, could penetrate to truths about the universe that no amount of casual observation would reveal.

The Foundation for What's Next

By 1915, Einstein had laid the groundwork for understanding phenomena that wouldn't be directly observed for decades: black holes, gravitational waves, the expansion of the universe, and the intimate connection between gravity and the flow of time. His insights provided the theoretical framework that would eventually explain some of the most extreme and beautiful phenomena in the cosmos.

But perhaps most importantly for our story, Einstein's revolutionary insights established that time dilation—the stretching and compression of time itself—is not just a mathematical curiosity but a fundamental feature of the universe. Every prediction of special and general relativity involving the behavior of time has been confirmed by experiment, often with extraordinary precision.

As we prepare to explore the specific mechanisms of time dilation, we carry with us Einstein's profound realization: the universe is far stranger and more wonderful than our everyday experience suggests, but it operates according to principles that human reason can discover and understand. The clockmaker's dilemma that puzzled nineteenth-century physics had been resolved, but the solution revealed that time itself is far more malleable and mysterious than anyone had dared to imagine.

Einstein's revolutionary journey from a teenage thought experiment to the theory of general relativity shows us that the deepest truths about nature often require us to abandon our most cherished assumptions. In learning to think like Einstein, we prepare ourselves to understand how speed and gravity can literally bend time itself—a phenomenon that transforms not just physics but our entire understanding of what it means to exist in this strange and beautiful cosmos.

Chapter 3: The Speed Limit of the Universe

Imagine, for a moment, that you're driving down an endless highway where every car, motorcycle, bicycle, and even every airplane overhead is bound by the same universal speed limit. No matter how powerful your engine, no matter how much fuel you carry, no matter how desperately you need to arrive at your destination, you can never exceed this cosmic speed limit. Moreover, this isn't just any ordinary speed limit—it's woven into the very fabric of reality itself, determining not just how fast things can move, but how space and time behave throughout the entire universe.

This is precisely the situation we find ourselves in as inhabitants of Einstein's universe, where the speed of light—approximately 299,792,458 meters per second—serves as nature's ultimate speed limit. But unlike the posted limits on our highways, this cosmic speed limit cannot be broken, bent, or ignored. It is as fundamental to the structure of reality as the existence of matter and energy themselves.

What makes this speed limit so special? Why does the universe seem to have chosen this particular number, and not something faster or slower? And perhaps most remarkably, how does this seemingly simple fact about light lead to the extraordinary phenomenon of time dilation? These questions take us to the heart of what makes Einstein's universe so strange and wonderful.

The Universal Constant

To understand why the speed of light is so special, we must first appreciate just how constant it really is. When we say that light travels at 299,792,458 meters per second, we're not describing an average or an approximation—we're stating a fundamental property of the universe that appears to be the same everywhere and at all times.

This constancy is not just impressive; it's almost incomprehensible in its precision. Observations of light from the most distant galaxies, emitted when the universe was less than a billion years old, show that light traveled at exactly the same speed then as it does today. The cosmic microwave background—the afterglow of the Big Bang itself—confirms that light has maintained this same velocity for over 13.8 billion years.

But the constancy of light's speed becomes even more remarkable when we consider that it remains the same regardless of how we observe it. Whether you're standing still, riding in a car, flying in an airplane, or even traveling on a spacecraft at tremendous velocity, you will always measure light to travel at exactly the same speed. This is fundamentally different from our everyday experience with moving objects.

If you're standing on a platform and a train passes by at 60 miles per hour, then throw a ball forward from the train at 30 miles per hour, the ball appears to move at 90 miles per hour relative to the platform. But light doesn't behave this way. If that same train were somehow traveling at half the speed of light and someone turned on a flashlight pointing forward, you would still measure that light traveling at exactly 299,792,458 meters per second—not at light speed plus the train's velocity.

This bizarre behavior of light was first confirmed by the famous Michelson-Morley experiment in 1887, which attempted to detect variations in light's speed as Earth moved through space. Despite the most careful measurements, no variation was ever found. Light's speed remained stubbornly constant, defying all expectations based on classical physics.

The Nature of the Limit

Why can nothing exceed the speed of light? The answer lies not in the properties of light itself, but in the fundamental structure of spacetime. Light doesn't move at this speed because it's special—rather, this speed is special because it represents the maximum rate at which any cause can produce an effect, any information can travel, or any influence can propagate through the universe.

To understand this, consider what would happen if faster-than-light travel were possible. Imagine you could send a signal from Earth to a distant star faster than light could make the same journey. Due to the relativity of simultaneity—the fact that different observers disagree about which events happen "at the same time"—there would exist reference frames in which your faster-than-light signal would arrive before it was sent.

This isn't just a mathematical curiosity; it would fundamentally violate causality—the principle that causes must precede their effects. In a universe where faster-than-light travel was possible, you could theoretically send messages into the past, creating paradoxes that would undermine the logical structure of physical law.

The speed of light, therefore, isn't really about light at all. It's the speed of spacetime itself—the rate at which the fabric of reality allows information and influence to propagate. Light happens to travel at this speed because photons are massless, but the speed itself is a property of the universe's geometry.

This cosmic speed limit affects everything in the universe, not just light. All massless particles must travel at exactly this speed, while all particles with mass must travel slower than this speed. The speed of light represents the boundary between what is possible and what is forbidden in Einstein's universe.

The Mathematics of the Limit

The relationship between mass, energy, and the cosmic speed limit is captured in one of the most important equations in physics. When we examine how energy relates to momentum for a moving object, we discover something remarkable: $\mathbf{E^2=(pc)^2+(mc^2)^2}$, where E is energy, p is momentum, m is mass, and c is the speed of light.

For a massless particle like a photon, this equation simplifies to $E=pc$, which means the particle must travel at the speed of light. For particles with mass, however, the situation is very different. As such a particle approaches the speed of light, its energy approaches infinity—meaning it would require infinite energy to actually reach light speed.

This isn't a limitation of our technology or understanding; it's a fundamental feature of the universe. The cosmic speed limit is enforced by the structure of spacetime itself. As objects approach this limit, time dilation and length contraction become more extreme, while the energy required for further acceleration grows without bound.

The mathematical factor that governs these relativistic effects is called the Lorentz factor: $\gamma = 1 / \sqrt{(1 - v^2/c^2)}$, where v is the object's velocity. This factor determines how much time dilates, how much lengths contract, and how energy increases with velocity.

At everyday speeds, this factor is essentially equal to 1, which is why we don't notice relativistic effects in daily life. But as speeds approach the cosmic limit, the Lorentz factor grows dramatically. At 86.6% of light speed, $\gamma=2$, meaning time passes half as slowly for the moving object. At 99.5% of light speed, $\gamma=10$, and time dilation becomes extreme. As velocity approaches the speed of light, γ approaches infinity.

The Fabric of Spacetime

The speed of light's role as the universe's speed limit reveals something profound about the nature of space and time. In Newton's universe, space and time were separate, absolute entities. Objects moved through space as time flowed uniformly for everyone. But Einstein's insights showed that space and time are woven together into a single fabric—spacetime—and this fabric has a structure that determines how events unfold.

The cosmic speed limit is built into this structure. Just as the geometry of a sphere determines that its surface curves back on itself, the geometry of spacetime determines that there exists a maximum speed for the propagation of cause and effect. This isn't an arbitrary rule imposed from outside; it's an inherent property of the fabric of reality.

When we think of spacetime this way, light's behavior becomes much less mysterious. Light doesn't "know" to travel at a constant speed; rather, massless particles naturally follow the straightest possible paths through spacetime, and the geometry of spacetime is such that these paths always correspond to the same speed when measured by any observer.

This geometric perspective also helps explain why nothing can exceed the speed of light. In spacetime, there exists a boundary between time-like trajectories (which connect cause to effect and can be followed by particles with mass) and spacelike trajectories (which would connect effect to cause and are forbidden to particles). The speed of light defines this boundary.

Practical Consequences

The universality of the cosmic speed limit has profound practical consequences that extend far beyond theoretical physics. Every GPS satellite orbiting Earth must account for the fact that its high-speed motion causes its onboard clocks to run differently than clocks on Earth's surface. Without corrections based on Einstein's theories, GPS systems would accumulate errors of several miles per day.

In particle accelerators, scientists routinely accelerate subatomic particles to velocities exceeding 99% of the speed of light. These experiments provide daily confirmation that the cosmic speed limit is real and absolute. No matter how much energy physicists pour into accelerating particles, they can never quite reach light speed, and the time dilation effects become exactly as Einstein predicted.

The cosmic speed limit also fundamentally shapes our understanding of the universe's structure. The finite speed of light means that when we look at distant galaxies, we see them as they were in the past. The observable universe—the sphere of space from which light has had time to reach us since the Big Bang—has a finite size determined by the age of the universe and the speed of light.

Even more remarkably, the speed of light determines the causal structure of the universe. Events can only influence each other if they are connected by trajectories that respect the cosmic speed limit. This creates a web of causal relationships that shapes everything from the formation of galaxies to the behavior of quantum particles.

The Fine Structure Constant

The specific value of the speed of light—299,792,458 meters per second—might seem arbitrary, but it's intimately connected to other fundamental properties of the universe. The speed of light appears in the fine structure constant, a dimensionless number approximately equal to 1/137 that characterizes the strength of electromagnetic interactions.

This constant appears throughout physics, from the behavior of atoms to the properties of light itself. While we can choose our units to make the speed of light any number we want, the fine structure constant remains the same regardless of units. This suggests that the speed of light's value reflects deep geometric properties of spacetime that we don't yet fully understand.

The fact that the fine structure constant has the value it does—and not some other value—remains one of the great mysteries of physics. If it were significantly different, atoms might not be stable, stars might not shine, and the universe as we know it might not exist. The cosmic speed limit appears to be fine-tuned for the existence of complexity and life.

Light in Matter

While the speed of light in vacuum is absolute and unchanging, light behaves differently when traveling through matter. In glass, water, or other transparent materials, light appears to slow down. But this apparent slowdown doesn't violate the cosmic speed limit—it's the result of complex interactions between light and matter.

At the microscopic level, photons still travel at the full speed of light between atoms, but they are constantly absorbed and re-emitted by the material's electrons. This process creates an effective delay that makes the light appear to propagate more slowly through the medium. The fundamental speed limit remains inviolate; only the bulk propagation through matter is affected.

This distinction is crucial for understanding how the cosmic speed limit works. The speed limit applies to the propagation of information and causal influence, not necessarily to the bulk motion of light through complex media. Even in exotic materials that can slow light to a crawl, no information travels faster than the fundamental limit.

Implications for Time

The cosmic speed limit is intimately connected to the nature of time itself. Because the speed of light is the same for all observers, and because space and time are linked in spacetime, observers moving at different velocities must experience time differently. This is the foundation of time dilation.

When an object moves at high speed, it's not just that its clocks run slowly—time itself passes more slowly for that object relative to a stationary observer. This isn't a mechanical effect on clocks; it's a fundamental feature of how time works in a universe with a cosmic speed limit.

The mathematical relationship is precise: time dilation is determined by how close an object's velocity comes to the cosmic speed limit. The closer to light speed, the more extreme the time dilation becomes. At the speed limit itself, time dilation becomes infinite—time would effectively stop from the perspective of outside observers.

This reveals that the cosmic speed limit isn't just a constraint on motion through space; it's a fundamental aspect of the structure of time itself. The speed of light sets the scale for how space and time trade off against each other in the fabric of spacetime.

The Universal Framework

The cosmic speed limit provides a universal framework that governs all physical processes. From the smallest quantum interactions to the largest cosmological structures, everything operates within the constraints imposed by this fundamental constant. It determines how quickly information can travel, how energy and momentum relate to each other, and how space and time are woven together.

This universality is part of what makes Einstein's relativity so powerful. Unlike Newton's mechanics, which required different rules for different situations, Einstein's framework provides a single, unified description that works everywhere in the universe. The cosmic speed limit is the key to this unification.

Understanding this speed limit as a fundamental feature of spacetime, rather than just a property of light, opens the door to comprehending the full range of relativistic effects. Time dilation, length contraction, mass-energy equivalence, and the curvature of spacetime all follow from the same underlying principle: the universe has a built-in speed limit that shapes the behavior of space and time themselves.

Looking Ahead

As we've seen, the cosmic speed limit is far more than just a fast speed—it's a fundamental constant that determines the structure of reality itself. This speed limit enforces causality, shapes the geometry of spacetime, and creates the conditions under which time dilation occurs.

But we've only begun to explore the implications of this cosmic speed limit. In the next chapter, we'll see how motion at speeds approaching this limit creates one of the most remarkable phenomena in physics: the stretching and compression of time itself. We'll discover that when objects move at relativistic speeds, time doesn't just appear to slow down—it actually does slow down, with consequences that seem to violate our deepest intuitions about the nature of temporal reality.

The cosmic speed limit isn't just a barrier to fast travel; it's the key to understanding how time itself can be stretched, compressed, and manipulated through motion. As we'll discover, racing against time takes on an entirely new meaning when time itself becomes elastic and malleable, responding to velocity in ways that reveal the true strangeness and beauty

PART II:

TIME DILATION THROUGH MOTION

Chapter 4: Racing Against Time

In 1971, two physicists named Joseph Hafele and Richard Keating did something that would have seemed like pure science fiction just decades earlier: they bought airline tickets around the world, not for vacation, but to test one of the most mind-bending predictions of Einstein's theory of relativity. They carried with them four of the most precise timekeeping devices ever created—cesium atomic clocks capable of measuring time to within billionths of a second. Their mission was audacious in its simplicity: to literally race against time itself and prove that moving clocks actually run slower than stationary ones.

What they discovered during their globe-circling flights would confirm something so counterintuitive that it challenges our most basic assumptions about the nature of reality. When they compared their flying clocks to identical clocks that had remained stationary on the ground, they found exactly what Einstein had predicted: the moving clocks had indeed lost time. Not because of mechanical failure or measurement error, but because time itself had actually passed more slowly for the clocks in motion.

This wasn't just a clever physics experiment—it was proof that time is not the unchanging, absolute entity that Newton had envisioned. Instead, time is elastic, malleable, and intimately connected to motion through space. When we race against time by moving very fast, time itself changes in response, stretching and compressing in ways that reveal the true strangeness of Einstein's universe.

The Foundation of Time Dilation

To understand how motion can affect the passage of time, we must return to the cosmic speed limit we explored in the previous chapter. Remember that the speed of light is constant for all observers, regardless of their own motion. This seemingly simple fact has profound consequences that ripple through every aspect of how space and time behave.

Consider what happens when you're traveling at high speed and you turn on a flashlight. The light beam travels away from you at exactly 299,792,458 meters per second, just as it would if you were standing still. But here's where things get strange: an observer watching you from the side also measures that same light beam traveling at exactly the same speed, even though you and your flashlight are moving rapidly relative to that observer.

How is this possible? In Newton's universe, velocities should simply add together. If you're moving at 100 kilometers per hour and throw a ball forward at 50 kilometers per hour, the ball moves at 150 kilometers per hour relative to a stationary observer. But light doesn't follow this rule—it maintains its constant speed regardless of the motion of its source or observer.

The only way this can work is if space and time themselves adjust to preserve the constancy of light speed. When you move at high velocity, space contracts in the direction of motion and time dilates—stretches out—relative to stationary observers. These aren't just mathematical tricks; they represent real, measurable changes in the fabric of spacetime itself.

The Mathematics of Moving Time

The relationship between velocity and time dilation is governed by the Lorentz factor, which we encountered briefly in the previous chapter: $\gamma = 1 / \sqrt{(1 - v^2/c^2)}$. This deceptively simple expression captures one of the most profound insights in all of physics: how time responds to motion through space.

When an object moves slowly compared to the speed of light, the velocity v is much smaller than c, making v^2/c^2 very small. In this case, the Lorentz factor is approximately equal to 1, and time dilation effects are negligible. This is why we don't notice relativistic effects in our daily lives—we simply don't move fast enough relative to the cosmic speed limit.

But as velocities approach the speed of light, the Lorentz factor grows dramatically. At 50% of light speed, γ=1.15, meaning time passes about 15% more slowly for the moving object. At 86.6% of light speed, γ=2, and time dilation becomes extreme—time passes half as slowly for the moving object compared to a stationary observer.

At 99% of light speed, γ=7.09, meaning that for every year that passes for a stationary observer, only about 51 days pass for the rapidly moving object. At 99.9% of light speed, γ=22.4, and the time dilation becomes almost science-fictional in its extremeness.

The mathematical relationship reveals something crucial: time dilation is not just a high-speed phenomenon. It occurs at any velocity greater than zero, but the effects are only noticeable at speeds that represent a significant fraction of the cosmic speed limit.

The Twin Paradox

Perhaps no thought experiment illustrates the reality of time dilation more dramatically than the famous twin paradox. Imagine identical twins, Alice and Bob, both 20 years old. Alice decides to take a journey to a nearby star system 4 light-years away, traveling at 86.6% of the speed of light. Bob stays on Earth.

From Bob's perspective on Earth, Alice's journey takes about 9.2 years—4.6 years to reach the destination and 4.6 years to return. During this time, Bob ages normally, so he's about 29 years old when Alice returns. But what about Alice?

Due to time dilation, Alice experiences time passing at half the normal rate during her high-speed journey. The Lorentz factor at 86.6% of light speed is 2, so Alice ages only 4.6 years during her 9.2-year journey. When she reunites with Bob, she's only 24.6 years old while Bob is 29 years old. Alice has literally traveled into the future by racing through space at relativistic speeds.

This isn't science fiction—it's a real consequence of the structure of spacetime. The effect becomes more dramatic as speeds approach the cosmic speed limit. If Alice could somehow travel at 99.5% of light speed (with γ=10), she would age only about one year during Bob's decade of waiting, returning to find her twin brother nine years older than herself.

The twin paradox reveals something profound about the nature of time: it's not a universal backdrop against which events unfold, but rather a local phenomenon that depends on motion through spacetime. Each observer carries their own personal time, and these personal times can diverge dramatically when relative velocities become large.

Real-World Evidence

The Hafele-Keating experiment provided the first direct measurement of time dilation using everyday objects—commercial airliners and atomic clocks. When they flew four clocks eastward around the world, the clocks lost about 59 nanoseconds compared to reference clocks on the ground. When they flew four clocks westward, the clocks gained about 273 nanoseconds.

These tiny time differences—measured in billionths of a second—might seem insignificant, but they matched Einstein's predictions with remarkable precision. The eastward-flying clocks experienced time dilation due to their increased speed relative to Earth's surface (since they were moving in the same direction as Earth's rotation). The westward-flying clocks, moving against Earth's rotation, had a lower effective speed and thus experienced less time dilation.

The experiment also had to account for gravitational effects, since the flying clocks were at a higher altitude where Earth's gravitational field is slightly weaker. We'll explore gravitational time dilation in later chapters, but the key point is that when both effects were properly calculated, the observed time differences matched Einstein's predictions to within about 10%.

Modern experiments have pushed the precision of time dilation measurements far beyond what Hafele and Keating achieved. In particle accelerators, scientists routinely observe time dilation effects with subatomic particles traveling at velocities exceeding 99% of the speed of light. The lifetimes of these fast-moving particles are extended exactly as Einstein's equations predict.

Muons from Space

One of the most elegant confirmations of time dilation comes from cosmic ray muons—subatomic particles created when high-energy cosmic rays collide with atoms in Earth's upper atmosphere. These muons are inherently unstable, with a natural lifetime of only 2.2 microseconds when at rest.

The problem is that muons created 15 kilometers above Earth's surface would normally decay long before reaching the ground, even if they traveled at the speed of light. At light speed, it would take 50 microseconds to travel 15 kilometers—more than 20 times longer than a muon's natural lifetime. Yet sensitive detectors at Earth's surface routinely detect cosmic ray muons.

The resolution lies in time dilation. The cosmic ray muons are traveling at velocities very close to the speed of light—typically around 99.8% of ***c***. At these extreme velocities, the Lorentz factor is approximately 16, meaning that the muons' internal clocks run 16 times slower than laboratory clocks. From our perspective, the muons live 16 times longer than they would if stationary, giving them enough time to reach Earth's surface before decaying.

This natural experiment provides daily confirmation of time dilation. Every cosmic ray muon that reaches a detector on Earth's surface is proof that time really does slow down for objects moving at relativistic speeds.

Particle Accelerator Precision

Modern particle accelerators provide even more precise tests of time dilation. In a 2014 experiment, physicists studied lithium ions traveling at 33.8% of the speed of light in a particle accelerator. They measured how long it took excited electrons in these ions to return to their ground state, comparing the results to identical measurements of stationary lithium ions.

The results were exactly what Einstein's theory predicted: the time interval for the excited electrons to return to ground state was longer for the moving ions than for the stationary ones, by precisely the amount calculated using the Lorentz factor. This type of experiment can now measure time dilation effects with extraordinary precision, confirming Einstein's predictions to many decimal places.

Even more remarkably, modern atomic clocks are now so precise that they can detect time dilation effects at much lower speeds. Scientists have measured time dilation for objects moving as slowly as a cyclist, and for elevation differences as small as 33 centimeters. These experiments reveal that time dilation is not just an exotic high-speed phenomenon—it's a fundamental feature of spacetime that affects everything in motion.

The Reality of Time Dilation

It's crucial to understand that time dilation is not an illusion or a measurement artifact—it represents a real difference in the passage of time itself. When the Hafele-Keating clocks showed different times after their round-the-world flights, it wasn't because the clocks had malfunctioned or because the experimenters had made errors. Time had literally passed at different rates for the moving and stationary clocks.

This reality has profound implications. If you could somehow observe a rapidly moving clock through a powerful telescope, you would see it ticking more slowly than an identical clock beside you. But here's the remarkable part: the person carrying that moving clock would observe their own clock ticking normally. From their perspective, it's your clock that appears to be running fast.

This reciprocity is a key feature of special relativity. Each observer sees the other's clock running slowly, and both observations are equally valid. There's no absolute reference frame that determines which clock is "really" slower—the effect depends entirely on the relative motion between observers.

However, the twin paradox resolves this apparent contradiction. When Alice returns from her high-speed journey, she and Bob can directly compare their ages, and Alice will indeed be younger. The asymmetry arises because Alice had to accelerate and decelerate during her journey, breaking the symmetry of uniform motion that characterizes special relativity.

Everyday Relativistic Technology

The reality of time dilation has immediate practical consequences in our technology-dependent world. The Global Positioning System (GPS) provides perhaps the most familiar example of relativistic effects in daily use. GPS satellites orbit Earth at speeds of about 14,000 kilometers per hour and at altitudes where Earth's gravitational field is weaker than at the surface.

Both effects influence the rate at which time passes for the satellites compared to receivers on Earth. The orbital motion causes time dilation that makes the satellite clocks run about 7 microseconds per day slower than Earth-based clocks. Meanwhile, the weaker gravity at orbital altitude causes the satellite clocks to run about 45 microseconds per day faster than Earth-based clocks.

The net effect is that satellite clocks run about 38 microseconds per day faster than Earth clocks. This might seem insignificant, but GPS requires extremely precise timing to determine positions accurately. Light travels about 30 centimeters in one nanosecond, so a timing error of 38 microseconds per day would cause GPS positions to drift by more than 11 kilometers per day without relativistic corrections.

Every time you use GPS navigation in your car or smartphone, you're relying on Einstein's equations to correct for time dilation effects. The engineers who designed GPS systems had to program Einstein's formulas directly into the system's software to maintain accuracy. Without these relativistic corrections, GPS would be useless for navigation within hours of being turned on.

The Relativity of Simultaneity

Time dilation is intimately connected to another counterintuitive aspect of Einstein's universe: the relativity of simultaneity. Events that appear simultaneous to one observer may not be simultaneous to another observer moving at a different velocity.

Imagine Alice and Bob are standing at opposite ends of a train platform when lightning strikes both ends simultaneously, as measured by an observer on the platform. But what does a passenger moving through the train station at high speed observe?

Due to the finite speed of light and the passenger's motion, the passenger will see one lightning strike before the other. The strikes that were simultaneous for the platform observer are not simultaneous for the moving passenger. This isn't just about when the light from the strikes reaches each observer—it reflects a fundamental property of spacetime itself.

The relativity of simultaneity reveals that there is no universal "now" that applies throughout the universe. Instead, each observer's notion of simultaneous events depends on their state of motion. This has profound implications for our understanding of time and causality, and it's intimately connected to the time dilation effects we experience when racing through space.

Energy and Time Dilation

As objects approach the cosmic speed limit, something remarkable happens to their energy. Einstein showed that the total energy of a moving object is given by $\mathbf{E=\gamma mc^2}$, where $\boldsymbol{\gamma}$ is the familiar Lorentz factor. As velocity increases, the Lorentz factor grows, meaning that the object's energy increases without bound.

This relationship reveals why nothing with mass can reach the speed of light: it would require infinite energy. But it also shows how time dilation is connected to the fundamental structure of matter and energy. The same factor that determines how much time dilates also determines how energy increases with velocity.

This connection isn't coincidental—it reflects the deep unity of space, time, matter, and energy in Einstein's universe. When we race against time by moving at high speeds, we're not just affecting the passage of time; we're also changing the fundamental energy content of matter itself.

Implications for Space Travel

The reality of time dilation has profound implications for future space exploration. If humans ever develop the technology to travel at significant fractions of the speed of light, time dilation effects would become a major factor in mission planning.

Consider a hypothetical mission to Proxima Centauri, our nearest stellar neighbor at 4.2 light-years away. If a spacecraft could maintain 90% of light speed, the journey would take about 4.7 years as measured by Earth-based observers. But due to time dilation, with γ=2.29, the travelers would experience only about 2.05 years during the outbound journey.

The return trip would involve similar time dilation, so the round-trip mission would take about 9.4 years as measured on Earth, but only about 4.1 years as experienced by the travelers. The crew would return to find that Earth had aged more than twice as much as they had during the mission.

At even higher speeds, the effects become more dramatic. At 99% of light speed, a round trip to Proxima Centauri would take about 8.4 years for Earth but only about 1.2 years for the travelers. The crew would return to find that seven years had passed on Earth for every year they experienced during their journey.

The Philosophical Implications

Time dilation challenges our most fundamental intuitions about the nature of time and existence. We're accustomed to thinking of time as universal—the same for everyone, everywhere, always flowing at the same steady rate. But Einstein's universe reveals a different reality: time is personal, local, and dependent on motion through space.

This has profound philosophical implications. If two observers can disagree about the duration between events, what does this mean for our understanding of causality, free will, and the nature of existence itself? How do we reconcile our subjective experience of a universal "now" with the objective reality that no such universal time exists?

These questions touch on some of the deepest issues in philosophy and physics. They reveal that time is not just a backdrop against which physical events unfold, but an active participant in the cosmic drama, shaping and being shaped by the matter and energy that fill the universe.

The Gateway to Greater Mysteries

The time dilation effects we've explored in this chapter—those arising from motion through space—represent just one aspect of time's malleable nature. As remarkable as these effects are, they pale in comparison to what happens when gravity enters the picture.

Einstein's 1915 theory of general relativity revealed that gravity itself can warp time, creating time dilation effects that can be far more extreme than those produced by motion alone. Near the surface of a neutron star, time passes noticeably slower than on Earth. At the event horizon of a black hole, time dilation becomes infinite from the perspective of outside observers.

But before we explore these gravitational effects on time, we must first understand the broader context of special relativity and its verified predictions. The experiments and observations we've discussed—from flying atomic clocks around the world to observing cosmic ray muons—have confirmed Einstein's special theory of relativity to extraordinary precision.

Every GPS satellite, every particle accelerator experiment, and every cosmic ray detection provides fresh confirmation that time really is elastic and malleable. When we race against time by moving through space, time itself responds, stretching and compressing in ways that reveal the true strangeness and beauty of the universe we inhabit.

The clockmaker's dilemma that we encountered in our first chapter has been definitively resolved. Time is not absolute and universal, but relative and personal. Our clocks don't measure some abstract, absolute time—they measure the local passage of time in their own reference frames, and these local times can differ dramatically depending on motion through space.

As we prepare to explore the even more dramatic effects of gravity on time, we carry with us the profound realization that racing against time is not just a metaphor—it's a literal description of how motion through space affects the passage of time itself. In Einstein's universe, when we move fast enough, we can actually win the race against time, arriving in the future younger than we would have been if we had stayed at rest.

The universe is far stranger and more wonderful than our everyday experience suggests, but it operates according to principles that human reason can discover, test, and verify. The flying clocks of Hafele and Keating opened a window into this strange universe, revealing that time itself is one of the most remarkable and malleable aspects of physical reality.

Chapter 5: The Mathematics of Motion

Maria Gonzalez stared at the equation written on her whiteboard in the particle physics laboratory at CERN, the world's largest particle accelerator. After months of analyzing data from high-speed proton collisions, she had just confirmed something remarkable: the mathematical formula that Einstein had derived over a century ago predicted, with breathtaking precision, exactly what her detectors had measured. Protons racing around CERN's 27-kilometer underground ring at 99.9999991% of the speed of light were behaving exactly as Einstein's mathematics dictated they should.

The formula on Maria's whiteboard looked deceptively simple: $\Delta t' = \gamma \Delta t$, where $\gamma = 1/\sqrt{(1 - v^2/c^2)}$. Yet this elegant mathematical relationship captures one of the most profound insights in all of science—that time itself responds to motion through space in ways that can be calculated with mathematical precision.

What makes Einstein's mathematics so remarkable is not just its accuracy, but its accessibility. Unlike many areas of advanced physics that require years of specialized study to comprehend, the essential mathematics of time dilation can be understood by anyone comfortable with basic algebra and square roots. The formulas are windows into the deepest truths about the nature of reality, yet they're simple enough that we can use them to calculate exactly how much time would slow down during a hypothetical journey to the stars.

The Heart of Time Dilation

At the center of everything we've discovered about time dilation lies a single mathematical expression: the Lorentz factor, named after the Dutch physicist Hendrik Lorentz who first derived it in the context of electromagnetic theory. This factor, represented by the Greek letter gamma (*γ*), determines how dramatically time, space, and mass change for objects in motion.

The Lorentz factor is defined as:

$$\gamma = \frac{1}{\sqrt{1 - \frac{v^2}{c^2}}}$$

Where ***v*** is the velocity of the moving object and *c* is the speed of light. This formula might look intimidating at first glance, but it becomes remarkably intuitive once we understand what each part represents.

The term v^2/c^2 is simply the square of the ratio between the object's speed and the cosmic speed limit. When an object moves slowly compared to light speed, this ratio is very small, making *γ* approximately equal to 1. But as the object's velocity approaches the speed of light, this ratio approaches 1, causing the denominator of the Lorentz factor to approach zero—and therefore causing *γ* to approach infinity.

This mathematical behavior captures the physical reality that nothing with mass can reach the speed of light.
As *γ* approaches infinity, all the relativistic effects—time dilation, length contraction, and mass increase—become infinitely large, requiring infinite energy to achieve.

Calculating Time Dilation

The relationship between the Lorentz factor and time dilation is beautifully direct. If Δt represents a time interval measured by a stationary observer, then the corresponding time interval Δt′ measured by an observer moving at velocity *v* is given by:

$$\Delta t' = \Delta t \gamma$$

This formula tells us that time intervals are shorter for the moving observer—time literally passes more slowly when you're in motion relative to a stationary reference frame.

Let's work through some concrete examples to see how dramatic these effects can become. Consider a spacecraft traveling at 50% of the speed of light, or ***v***=0.5***c***. To calculate the Lorentz factor:

$$\boldsymbol{\gamma} = 1/\sqrt{(1-\{0.5^2/1^2\})} = 1/\sqrt{(1-.25)} = 1/\sqrt{(.75)} = 1/0.866= 1.155$$

This means that for every hour that passes for a stationary observer, only about 52 minutes pass for someone traveling at half the speed of light. The effect is noticeable but not yet dramatic.

Now consider a spacecraft traveling at 90% of light speed, or ***v***=0.9***c***:

$$\gamma == 1/\sqrt{(1-\{0.9^2/1^2\})}= 1/\sqrt{(1-.81)}= 1/\sqrt{(0.19)}= 1/0.4359= 2.294$$

At this speed, time dilation becomes significant. For every hour experienced by a stationary observer, the high-speed traveler experiences only about 26 minutes. A journey that takes 10 years from Earth's perspective would feel like only 4.4 years to the traveler.

The effects become truly dramatic as we approach the cosmic speed limit. At 99% of light speed:

$\gamma= 1/\sqrt{(1-\{0.99^2/1^2\})}= 1/\sqrt{(1-.9801)}= 1/\sqrt{(0.0199)}= 1/0.1411= 7.089$

Now the time dilation is extreme. A 10-year journey from Earth's perspective would feel like only 1.4 years to the traveler.

At 99.9% of light speed, $\gamma=22.37$, meaning time passes more than 22 times slower for the traveler. A decade-long journey would feel like less than 5 months.

The Behavior Near Light Speed

The mathematical behavior of the Lorentz factor reveals why the speed of light serves as an absolute barrier for objects with mass. As velocity approaches *cc*, the term v^2/c^2 approaches 1, making the denominator $1-v^2/c^2$ approach zero. When a denominator approaches zero, the fraction approaches infinity.

This mathematical singularity corresponds to a physical impossibility. Infinite time dilation would mean that time stops completely from the perspective of outside observers. Infinite energy would be required to reach such a state. The mathematics perfectly captures the physical reality that light speed is unreachable for massive objects.

We can see this behavior clearly by examining how γ changes for velocities very close to light speed:

- At 99.9% of c: $\gamma=22.37$
- At 99.99% of c: $\gamma=70.71$

- At 99.999% of *c*: $\gamma=223.6$
- At 99.9999% of *c*: $\gamma=707.1$

The pattern is clear: each additional "9" in the percentage multiplies γ by approximately $10\approx3.16$. The Lorentz factor grows without bound as velocity approaches light speed, reflecting the infinite energy barrier that prevents massive objects from reaching this cosmic speed limit.

Length Contraction Mathematics

Time dilation is intimately connected to another relativistic effect: length contraction. Objects moving at high speeds become shorter in the direction of motion, as measured by stationary observers. The mathematics of length contraction mirrors that of time dilation, but with an important difference.

If L_0 represents the proper length of an object (its length when at rest), then its contracted length L_L when moving at velocity v is:

$L=L_0\gamma$

Notice that while time intervals become longer for moving observers (time dilation), lengths become shorter. This reciprocal relationship between space and time reflects their deep interconnection in the fabric of spacetime.

Consider our spacecraft traveling at 90% of light speed, where $\gamma=2.294$. A 100-meter spacecraft would appear to be only about 43.6 meters long to a stationary observer. At 99% of light speed, with $\gamma=7.089$, the same spacecraft would appear to contract to just 14.1 meters in length.

This length contraction isn't a mechanical compression—it's a fundamental property of how space and time relate to each other. From the spacecraft's perspective, its own length remains unchanged, but distances in the outside universe appear contracted in the direction of motion.

The Relativistic Energy-Mass Relationship

Perhaps the most famous equation in all of physics emerges from the mathematical framework of special relativity: $E=mc^2$. But this familiar formula is actually just a special case of a more general relationship that applies to objects in motion.

The total energy of an object moving at velocity v is given by:

$$E=\gamma mc^2$$

When the object is at rest ($v=0$), the Lorentz factor $\gamma=1$, and we recover Einstein's famous equation $E=mc^2$. This rest energy represents the enormous amount of energy contained within matter itself, even when that matter is completely stationary.

But when objects are in motion, their total energy increases by the Lorentz factor. This additional energy is called kinetic energy, and it grows without bound as velocity approaches light speed. The mathematical relationship shows us why accelerating massive objects to light speed is impossible—it would require infinite energy.

We can separate the total energy into rest energy and kinetic energy:

$$KE=(\gamma-1)mc^2$$

For low velocities, this reduces to the familiar classical formula $KE=(1/2)mv^2$, but for high velocities, the relativistic effects become dominant. At 90% of light speed, with $\gamma=2.294$, an object's kinetic energy is $1.294mc^2$ - more than its entire rest mass energy.

The Relativistic Momentum Formula

Classical momentum is simply mass times velocity: $p=mv$. But in Einstein's universe, momentum must be modified to account for relativistic effects:

$$p=\gamma mv$$

This relativistic momentum approaches infinity as velocity approaches light speed, even for objects with finite mass. This mathematical behavior reflects the physical reality that infinite force would be required to accelerate a massive object to light speed.

The relationship between energy and momentum in relativity is given by:

$$E^2=(pc)^2+(mc^2)^2$$

This formula applies to all objects, regardless of their velocity. For massless particles like photons, $m=0$, so $E=pc$, confirming that light carries both energy and momentum despite having no mass.

Mathematical Symmetry and Physical Insight

One of the most beautiful aspects of Einstein's mathematics is its symmetry. The same Lorentz factor γ appears in the formulas for time dilation, length contraction, mass increase, and energy change. This mathematical unity reflects a deep physical truth: space, time, mass, and energy are not separate entities but different aspects of a unified spacetime geometry.

The transformation equations that connect measurements made by different observers—the Lorentz transformations—exhibit this same mathematical elegance. If two observers are moving relative to each other at velocity *v*, their space and time coordinates are related by:

$$x' = \gamma(x - vt) \qquad t' = \gamma(t - vx/c^2)$$

These equations show how space and time mix together when we change from one reference frame to another. What one observer measures as purely spatial separation, another observer measures as a mixture of spatial and temporal separation.

Working with Practical Examples

To make these mathematical relationships concrete, let's work through a detailed example involving interstellar travel. Imagine a mission to Proxima Centauri, 4.24 light-years away, traveling at 80% of light speed.

First, let's calculate the Lorentz factor:

$\gamma = 1/\sqrt{(1 - \{0.8^2/1^2\})} = 1/\sqrt{(1 - .64)} = 1/\sqrt{(0.36)} = 1/0.6 = 1.667$

From Earth's perspective, the journey to Proxima Centauri takes:

t = {4.24 light-years}{0.8c} = 5.3 years

But from the spacecraft's perspective, due to time dilation:

$t' = t\gamma$ = {5.3}{1.667} = 3.18 years

The spacecraft crew experiences the journey as taking only 3.18 years, while Earth observers measure it as taking 5.3 years. This 2.12-year difference represents real time that could be "saved" through relativistic travel.

From the spacecraft's perspective, the distance to Proxima Centauri appears contracted:

$L = 4.24\gamma = \{4.24\}\{1.667\} = 2.54$ light-years

This length contraction is consistent with the time dilation—the shorter distance and slower passage of time combine to maintain the same measured speed of light in both reference frames.

Energy Requirements for Relativistic Travel

The mathematical relationships also reveal the enormous energy requirements for high-speed space travel. Consider accelerating a 1,000-ton spacecraft (mass $m=10^6$ kg) to 80% of light speed.

The kinetic energy required is:

$KE = (\gamma-1)mc^2 = (1.667 - 1) \times 10^6 \times (3\times10^8)^2 = 6.0 \times 10^{22}$ joules

To put this in perspective, this is about 500 times the total annual energy consumption of human civilization. The mathematical formula reveals why relativistic space travel remains beyond our current technological capabilities.

The Mathematics of Particle Accelerators

Modern particle accelerators routinely demonstrate these mathematical relationships with extraordinary precision. At CERN's Large Hadron Collider, protons are accelerated to 99.9999991% of light speed, corresponding to $\gamma = 7,461$.

At these extreme velocities, the protons' kinetic energy is about 7,460 times their rest mass energy. The mathematical precision required to control and measure such effects pushes the boundaries of engineering and computation, yet the results consistently confirm Einstein's equations to many decimal places.

The mathematical framework allows physicists to predict exactly how particles will behave at these extreme speeds, enabling the discovery of new particles and the testing of fundamental theories. Every collision at CERN is a confirmation of the mathematical structure of spacetime.

Graphical Understanding

While the mathematical formulas capture the precise relationships, graphical representations help us visualize how dramatic relativistic effects become. The Lorentz factor γ remains close to 1 for velocities up to about 50% of light speed, then rises slowly until about 90% of light speed, after which it increases explosively toward infinity.

This behavior explains why relativistic effects are negligible in everyday experience but become dominant at the extreme velocities achieved in particle accelerators and astronomical phenomena. The mathematics reveals a universe with two distinct regimes: the familiar low-speed world of classical physics and the strange high-speed realm of relativity.

Mathematical Precision in GPS

The Global Positioning System provides a daily demonstration of relativistic mathematics in action. GPS satellites orbit at velocities of about 14,000 kilometers per hour, corresponding to $v/c = 1.3 \times 10^{-8}$. The resulting Lorentz factor is:

$$\gamma = 1.000000000085$$

This tiny deviation from 1 might seem insignificant, but it corresponds to a time dilation effect of about 7 microseconds per day. GPS requires timing precision much better than this, so the relativistic corrections are essential for accurate positioning.

The mathematical formulas allow engineers to calculate these corrections precisely and program them into GPS software. Without Einstein's equations, GPS would accumulate errors of several miles per day, making the system useless for navigation.

The Deep Meaning of Mathematical Beauty

Einstein once remarked that "the most incomprehensible thing about the universe is that it is comprehensible." The mathematical relationships we've explored embody this profound truth. Simple algebraic expressions capture the deepest aspects of space, time, and motion, revealing a universe governed by mathematical principles of extraordinary beauty and precision.

The fact that the same Lorentz factor appears in multiple physical phenomena—time dilation, length contraction, mass-energy equivalence—suggests a fundamental unity underlying the apparent diversity of physical effects. Mathematics serves not just as a tool for calculation but as a window into the geometric structure of reality itself.

This mathematical beauty extends beyond mere aesthetics. The elegant formulas suggest that the universe operates according to principles that human reason can discover and understand. The mathematical relationships don't just describe physical phenomena—they reveal why those phenomena must be exactly as they are, given the structure of spacetime.

Experimental Verification

Every mathematical prediction of special relativity has been confirmed by experiment with extraordinary precision. The time dilation formula correctly predicts the lifetimes of fast-moving particles, the behavior of atomic clocks in motion, and the operation of GPS satellites. Length contraction has been confirmed through observations of high-energy cosmic rays and particle accelerator experiments.

The energy-momentum relationships are verified millions of times per second in particle accelerators around the world. The mathematical framework allows physicists to design experiments with confidence that the results will match theoretical predictions, often to many decimal places.

This experimental confirmation transforms the mathematical formulas from abstract mathematical relationships into verified descriptions of physical reality. The equations don't just work mathematically—they accurately describe how the universe actually behaves.

Looking Forward

The mathematical relationships we've explored in this chapter provide the foundation for understanding even more exotic phenomena. When we consider the effects of gravity on spacetime, we'll encounter mathematical relationships that are more complex but equally beautiful. Einstein's field equations, which describe how matter and energy curve spacetime, represent the mathematical culmination of his insights about the nature of gravity and time.

But even these more advanced mathematical concepts build on the foundation we've established here. The Lorentz factor and the time dilation formula remain central to understanding all relativistic phenomena, from the behavior of GPS satellites to the extreme physics near black holes.

The mathematics of motion reveals that Einstein's universe is not just stranger than our everyday experience suggests—it's also more beautiful, more elegant, and more comprehensible than anyone had dared to imagine. The simple formulas we've explored encode profound truths about the nature of reality, truths that become accessible to anyone willing to engage with the mathematical language of the universe.

As we prepare to explore the gravitational effects on time that Einstein discovered in his general theory of relativity, we carry with us the profound realization that mathematics provides not just a tool for calculation but a pathway to understanding the deepest aspects of existence. The equations on Maria Gonzalez's whiteboard at CERN don't just predict experimental results—they reveal the mathematical harmony that underlies the cosmic dance of space, time, and motion.

Chapter 6: Living in a Relativistic World

Sarah Chen glanced at her smartphone as she stepped out of the coffee shop in downtown San Francisco, checking the GPS navigation app that would guide her through the city's maze of streets to her afternoon meeting. The little blue dot on her screen showed her precise location—accurate to within a few yards—while the app calculated the fastest route through traffic in real time. What Sarah didn't realize as she followed those turn-by-turn directions was that she was experiencing one of the most remarkable demonstrations of Einstein's relativity in everyday life.

Twenty thousand kilometers above her head, a constellation of 31 GPS satellites raced through space at speeds of nearly 14,000 kilometers per hour. Each satellite carried atomic clocks so precise they could measure time to within a few billionths of a second. But here's the extraordinary part: without constant corrections based on Einstein's equations for time dilation, Sarah's GPS would be useless within hours, accumulating errors that would make navigation impossible.

Every step Sarah took while following her GPS directions was guided by calculations that accounted for the fact that time passes differently for those fast-moving satellites than it does for people on Earth's surface. This is the reality of living in a relativistic world—Einstein's insights about space and time aren't just abstract theories discussed in physics classrooms, they're woven into the fabric of our daily technological lives.

The GPS Revolution

The Global Positioning System represents perhaps the most ubiquitous application of relativistic physics in human history. More than four billion people carry GPS-enabled devices, yet few realize that each position calculation requires Einstein's equations to work properly.

The fundamental principle behind GPS is elegantly simple: measure the time it takes radio signals to travel from satellites to your receiver, calculate the distance based on the speed of light, and use triangulation to determine your precise location. But executing this principle requires timing precision that pushes the boundaries of measurement—accurate to within nanoseconds over periods of hours and days.

Here's where relativity becomes essential. GPS satellites orbit Earth at an altitude of approximately 20,200 kilometers, moving at speeds of 3.87 kilometers per second relative to Earth's surface. At these velocities, special relativistic time dilation causes the satellite clocks to run about 7 microseconds per day slower than identical clocks on Earth.

But there's a complicating factor: the satellites are also in a weaker gravitational field than objects on Earth's surface. As we'll explore in detail in later chapters, gravitational time dilation causes clocks in weaker gravity to run faster. This gravitational effect causes the satellite clocks to gain about 45 microseconds per day relative to Earth-based clocks.

The combined effect is that GPS satellite clocks run about 38 microseconds per day faster than Earth clocks. This might seem insignificant, but remember that GPS requires extraordinary precision. Light travels about 30 centimeters in one nanosecond, so an error of 38 microseconds corresponds to a distance error of about 11 kilometers. Without relativistic corrections, GPS would become useless for navigation within a single day.

The solution requires programming Einstein's equations directly into the GPS system's software. Every GPS receiver on Earth—from the navigation system in your car to the location services in your smartphone—relies on continuous relativistic corrections to maintain accuracy. The satellites' onboard computers constantly adjust their clocks based on relativistic predictions, while ground-based control systems monitor and fine-tune these corrections.

This represents one of the most successful large-scale applications of theoretical physics in human history. Einstein's equations, derived from pure thought experiments about the nature of space and time, have become essential for a technology that affects billions of people daily.

The Electromagnetic Foundation

The relativistic effects that govern GPS satellites are just one example of how Einstein's insights shape our technological world. In fact, many of the electromagnetic technologies we take for granted depend on relativistic principles, even when the speeds involved are much lower than those of orbiting satellites.

Consider the simple electric generator that provides power to your home. When a wire moves through a magnetic field, it generates an electric current—this is the fundamental principle behind all electrical generation, from massive hydroelectric dams to small bicycle dynamos. But the mechanism by which this works is inherently relativistic.

From the perspective of the moving wire, the magnetic field appears as both magnetic and electric fields due to relativistic effects. The moving charged particles in the wire interact with these fields in ways that can only be fully understood through Einstein's framework. Even at the relatively low speeds of rotating generators—perhaps a few hundred meters per second at the tips of turbine blades—relativistic effects play a crucial role in determining exactly how much current is generated.

The same principles apply to electric motors, transformers, and countless other electromagnetic devices. While the effects are small at everyday speeds, they're precisely calculable using relativistic equations, and engineers must account for them when designing high-precision electrical systems.

The Color of Gold

One of the most surprising applications of relativistic physics appears in something as mundane as the color of gold jewelry. Gold's distinctive yellow color—which has made it prized throughout human history—is a direct consequence of relativistic effects on the atoms themselves.

Gold atoms contain 79 electrons, with the outermost electrons moving at speeds that represent a significant fraction of the speed of light. These high speeds cause the electron orbitals to contract relativistically, changing the energy levels at which electrons can exist within the atom. This orbital contraction affects exactly which wavelengths of light gold atoms can absorb and reflect.

Without relativistic effects, gold would appear silvery-white, similar to silver or aluminum. The warm yellow hue that makes gold so distinctive results from the same mathematical principles that govern GPS satellites and particle accelerators—Einstein's equations operating at the atomic scale.

Similar relativistic effects influence the properties of other heavy elements, including mercury's liquid state at room temperature and the chemical behavior of elements like tungsten and platinum. The entire bottom portion of the periodic table owes many of its properties to relativistic effects on electron behavior.

Television and Display Technology

The television screen or computer monitor you're reading this on may also demonstrate relativistic principles, depending on its age and technology. Cathode ray tube (CRT) displays, which dominated television and computer technology for decades, rely directly on relativistic corrections to produce clear images.

CRT displays work by firing electrons at high speeds toward a phosphorescent screen. These electrons are accelerated to velocities of 20-30% of the speed of light—fast enough that relativistic effects become significant. As the electrons approach these speeds, their effective mass increases according to Einstein's equations, changing how they respond to the magnetic fields used to steer them across the screen.

Without accounting for these relativistic effects, the magnetic deflection systems in CRT displays would be miscalibrated, causing images to appear distorted or out of focus.

Engineers designing CRT systems had to incorporate Einstein's equations into their calculations to ensure that electrons struck the screen at precisely the right locations to form clear images.

While newer LCD and OLED displays don't rely on high-speed electron beams, they still depend on electromagnetic principles that are fundamentally relativistic. The behavior of light within these displays—how it propagates through liquid crystals or interacts with organic compounds—follows laws that can only be fully understood through Einstein's framework.

Radar and Speed Detection

The radar gun that police officers use to measure vehicle speeds provides another everyday demonstration of relativistic physics. These devices work by emitting radio waves and detecting the frequency shift of the reflected signal—an effect known as the Doppler shift.

When radio waves reflect off a moving object, their frequency changes in a way that depends on the object's velocity. This frequency shift allows the radar gun to calculate the speed of the target vehicle. But the precise relationship between frequency shift and velocity comes from relativistic equations, not classical physics.

At the speeds of typical highway traffic—say, 100 kilometers per hour—the difference between classical and relativistic calculations is tiny, much less than the measurement precision of ordinary radar guns. But the underlying physics is relativistic, and for very precise measurements or very high speeds, the relativistic corrections become essential.

Military radar systems, which must track aircraft and missiles moving at supersonic speeds, require more sophisticated relativistic calculations. Air traffic control systems, which coordinate thousands of flights traveling at hundreds of kilometers per hour, rely on timing and positioning calculations that ultimately depend on Einstein's equations.

Industrial Applications

Modern manufacturing increasingly relies on technologies that demonstrate relativistic effects at the industrial scale. Electron beam machining (EBM), for instance, uses focused beams of electrons accelerated to 50-80% of the speed of light to cut precise patterns in metal and other materials.

At these extreme velocities, the electrons' kinetic energy must be calculated using relativistic equations rather than classical formulas. The energy transfer that occurs when these high-speed electrons strike the workpiece—melting or vaporizing material to create precise cuts—depends entirely on relativistic energy-momentum relationships.

EBM systems are used throughout the aerospace, automotive, and medical device industries to create components that require extraordinary precision. The cutting patterns are programmed using software that incorporates Einstein's equations to predict exactly how much energy the electron beam will deposit at each point along its path.

Similar principles apply to other high-energy manufacturing processes, including particle beam welding, ion implantation for semiconductor manufacturing, and various forms of radiation processing. These technologies represent the practical application of particle accelerator physics to industrial production.

Particle Physics in Daily Life

While most people never directly interact with particle accelerators, the technologies developed for high-energy physics research have found countless applications in daily life, all based on relativistic principles. Medical imaging systems, cancer treatment devices, and materials analysis equipment all trace their origins to particle physics research.

Magnetic Resonance Imaging (MRI) machines, for instance, rely on precise control of electromagnetic fields and radiofrequency pulses to image the interior of the human body. The physics governing these interactions—how atomic nuclei respond to electromagnetic radiation—is fundamentally relativistic, even though the particles involved move at relatively low speeds.

Positron Emission Tomography (PET) scanners used in medical diagnosis rely on the detection of antimatter particles—positrons—that are created by radioactive decay processes. The interaction between positrons and electrons, which produces the gamma rays that PET scanners detect, follows relativistic equations that determine exactly how much energy is released and in which directions.

Even smoke detectors in homes and offices often contain small amounts of radioactive material that undergoes decay processes governed by relativistic physics. The detection mechanism relies on the ionization effects of high-energy particles—effects that can only be properly calculated using Einstein's framework.

The Energy of Stars

Perhaps the most fundamental application of relativistic physics affects every moment of life on Earth: the nuclear fusion reactions that power the Sun. Einstein's famous equation $E=mc^2$ describes the conversion of matter into energy that makes stars shine and ultimately sustains all life on our planet.

In the Sun's core, hydrogen nuclei collide with such violence that they overcome their mutual electrical repulsion and fuse together to form helium. The helium nucleus has slightly less mass than the original hydrogen nuclei, and this "missing" mass is converted directly into energy according to Einstein's equation.

The amount of energy produced is staggering: the Sun converts about 4 million tons of matter into pure energy every second, releasing enough power to heat and light our planet across 150 million kilometers of space. This energy production has continued steadily for 4.6 billion years and will continue for several billion more, all because of the mass-energy equivalence discovered by Einstein.

Without this relativistic effect, stars could not generate the sustained energy output necessary to support planetary systems. The universe would be a cold, dark place where complex chemistry and life as we know it could never have developed. Every plant that grows in sunlight, every solar panel that generates electricity, and every weather pattern driven by solar heating demonstrates the practical consequences of Einstein's insights.

Magnetic Interactions

Even simple electromagnetic phenomena that we encounter daily, such as the attraction between magnets or the operation of electric motors, involve subtle relativistic effects. When electric currents flow through parallel wires, the wires experience magnetic forces that depend on relativistic length contraction effects.

Consider two parallel wires carrying electric current in the same direction. From the perspective of the moving electrons in one wire, the positive charges in the other wire appear more closely spaced due to relativistic length contraction, while the electrons in the other wire appear more spread out. This creates an apparent charge imbalance that results in magnetic attraction between the wires.

If the currents flow in opposite directions, the relativistic effects reverse, causing the wires to repel each other. These forces are precisely what allow electric motors to work, speakers to produce sound, and magnetic levitation systems to function. While the relativistic effects are tiny at the speeds of electrons in ordinary wires, they're fundamental to understanding why electromagnetism works as it does.

Precision Timekeeping

The atomic clocks used in GPS satellites represent just one example of how relativistic effects influence precision timekeeping. The most accurate atomic clocks on Earth—capable of maintaining accuracy to within one second over the entire age of the universe—must account for even subtler relativistic effects.

These ultra-precise clocks are sensitive enough to detect time dilation effects from moving just a few meters per second, or from changes in altitude of less than a meter. Scientists have used such clocks to measure time dilation effects in moving vehicles, in elevators, and even for people simply walking around a laboratory.

This precision enables applications far beyond navigation. Precise timekeeping is essential for internet synchronization, financial trading, power grid coordination, and scientific research. The global network of atomic clocks that maintains coordinated universal time relies on continuous relativistic corrections to keep the world's clocks synchronized to within nanoseconds.

The Future of Relativistic Technology

As technology continues to advance, relativistic effects are becoming important in an ever-widening range of applications. Modern particle accelerators routinely push particles to velocities exceeding 99.9% of the speed of light, requiring sophisticated relativistic calculations for their operation.

Future technologies may rely even more heavily on relativistic effects. Proposed fusion power systems would harness the same mass-energy conversion that powers the Sun. Advanced propulsion systems for spacecraft might achieve velocities high enough that time dilation effects become significant for human passengers.

Even in computing, relativistic effects are beginning to matter. As computer processors become faster and more compact, the electrons carrying information approach velocities where relativistic effects on their behavior become measurable. Future quantum computers may rely on relativistic effects on subatomic particles to perform calculations impossible with classical systems.

The Invisible Foundation

What makes living in a relativistic world so remarkable is how invisible these effects usually are in daily experience. The GPS in your phone works flawlessly without requiring you to understand time dilation. Your television displays clear images without you thinking about electron beam deflection calculations. The Sun shines steadily without you contemplating mass-energy conversion.

Yet all these technologies depend absolutely on the relativistic structure of spacetime that Einstein revealed. Remove the relativistic corrections from GPS, and navigation systems worldwide would fail within hours. Ignore relativistic effects in CRT manufacturing, and television images would be hopelessly distorted. Take away the mass-energy conversion in stellar cores, and the universe would be dark and lifeless.

This hidden foundation of relativistic physics extends throughout our technological civilization. From the smartphones in our pockets to the power grids that light our cities, from the medical devices that diagnose disease to the satellites that enable global communication, Einstein's insights about the nature of space and time have become woven into the fabric of modern life.

The Precision Revolution

Perhaps most remarkably, many of these applications require not just that Einstein's equations be correct, but that they be correct to extraordinary precision. GPS navigation accuracy depends on timing measurements precise to nanoseconds.

Particle accelerator experiments require energy calculations accurate to many decimal places. Atomic clocks must account for relativistic effects at the level of parts per trillion.

This precision represents one of the great triumphs of human knowledge. Theories developed through pure thought about the fundamental nature of reality have proven accurate enough to enable technologies that affect billions of people daily. The mathematical relationships Einstein derived from his insights about space and time have become the foundation for some of our most precise and essential technologies.

A Relativistic Civilization

As we've seen throughout this chapter, we don't just live in a universe governed by relativistic physics—we've built a technological civilization that depends absolutely on our understanding of these effects. From the atomic scale properties of matter to the orbital mechanics of satellites, from the nuclear reactions in stellar cores to the electromagnetic interactions in our devices, relativistic physics provides the invisible foundation that makes modern technology possible.

This represents a profound achievement of human understanding. In little more than a century since Einstein first proposed his theories, we've learned to harness relativistic effects for practical purposes that affect nearly every aspect of daily life. The young patent clerk who wondered about riding alongside a light beam could never have imagined that his insights would someday guide navigation systems for billions of people.

Yet this is only the beginning of our exploration of time dilation. The velocity-based effects we've examined in this chapter, remarkable as they are, represent just one way that time can be stretched and compressed. When gravity enters the picture, the effects become even more dramatic, revealing aspects of spacetime that challenge our deepest assumptions about the nature of reality itself.

As we prepare to explore gravitational time dilation, we carry with us the profound realization that Einstein's universe isn't just a theoretical curiosity—it's the world we actually live in, complete with GPS navigation, atomic clocks, and stellar fusion. The mathematical relationships that seemed so abstract in previous chapters have proven to be practical tools for building the technologies that define our modern world.

Living in a relativistic world means living in a universe where time itself is malleable, where space and time are woven together, and where the deepest truths about reality can be discovered through human reason and verified through daily experience. It's a universe that's stranger and more wonderful than common sense suggests, but also more comprehensible and more useful than anyone could have dared to hope.

PART III:
GRAVITY'S GRIP ON TIME

Chapter 7: The Equivalence Principle

Dr. Elena Vasquez stepped into the elevator on the 40th floor of the physics building at Stanford University, her mind still occupied with the lecture she had just delivered on Newton's laws of motion. As the elevator began its descent, she felt the familiar sensation of her stomach seeming to rise slightly—that moment of apparent weightlessness that occurs at the beginning of any downward journey. But this time, something clicked in her mind about that sensation, and she found herself thinking about one of the most profound insights in all of physics.

What if, Elena wondered, the elevator cable suddenly snapped? For those terrifying seconds of free fall, she would experience complete weightlessness. More remarkably, from her perspective inside the falling elevator, the laws of physics would be indistinguishable from those she would experience floating in the vacuum of deep space, far from any gravitational influence. This seemingly simple observation—that free fall eliminates the local effects of gravity—was the key insight that led Einstein to his most revolutionary discovery: the principle that would transform our understanding of gravity, space, time, and the very fabric of reality itself.

This is the equivalence principle, and it represents perhaps the most elegant and far-reaching insight in the history of physics. What began as Einstein's "happiest thought"—the realization that a person in free fall experiences no gravitational force—would ultimately reveal that gravity is not a force at all, but rather a manifestation of the curvature of spacetime itself. More importantly for our journey through the universe of time dilation, the equivalence principle opened the door to understanding how gravity can stretch and compress time in ways even more dramatic than the velocity effects we've explored.

The Newtonian Foundation

To appreciate the revolutionary nature of Einstein's insight, we must first understand the puzzle that had perplexed physicists for centuries. In Newton's universe, there appeared to be two completely different types of mass, with no obvious reason why they should be related.

Inertial mass determines how much an object resists acceleration when a force is applied to it. This is the mass that appears in Newton's second law: $F = ma$. If you push on a bowling ball and a tennis ball with the same force, the bowling ball accelerates more slowly because it has greater inertial mass. This type of mass is a measure of an object's "stubbornness"—its resistance to changes in motion.

Gravitational mass, on the other hand, determines how strongly an object responds to gravitational fields and how strongly it generates such fields. This is the mass that appears in Newton's law of universal gravitation: $F = G(m_1m_2/r^2)$. The gravitational mass determines how much gravitational force an object experiences and how much gravitational force it exerts on other objects.

Here's the puzzle: these two types of mass could, in principle, be completely unrelated. There's no obvious reason why the property that makes an object hard to accelerate should have anything to do with how it responds to gravity. Yet careful experiments, beginning with Galileo's legendary tests at the Leaning Tower of Pisa, showed that all objects fall at the same rate in a gravitational field, regardless of their composition or mass.

This strange coincidence—that inertial mass and gravitational mass are always exactly equal—was encoded in Newton's physics as an observed fact, but Newton himself admitted he couldn't explain why it should be true. It seemed like a remarkable accident of nature that these two fundamentally different quantities were always numerically identical.

Galileo's Revolutionary Discovery

The roots of the equivalence principle stretch back more than 400 years to Galileo Galilei's groundbreaking experiments with falling objects. According to Aristotelian physics, which had dominated scientific thinking for nearly 2,000 years, heavier objects should fall faster than lighter ones. This seemed intuitively obvious—surely a massive stone would plummet to earth more quickly than a feather.

But Galileo's careful experiments revealed something extraordinary: when air resistance is eliminated, all objects fall at exactly the same rate, regardless of their weight, size, or composition. A feather and a hammer dropped simultaneously in a vacuum will hit the ground at precisely the same time. This principle, known as the universality of free fall, was so counterintuitive that it took centuries for physicists to fully appreciate its profound implications.

The universality of free fall means that gravitational acceleration—9.8 meters per second squared near Earth's surface—is the same for all objects. But why should this be true? In Newton's framework, the gravitational force on an object is proportional to its gravitational mass: $F_{gravity} = m_{gravitational} \times g$.

Meanwhile, the acceleration that force produces depends on the object's inertial mass: $F = m_{inertial} \times a$.

For all objects to have the same gravitational acceleration, we need: $\{m_{gravitational}\}\{m_{inertial}\}$ = *constant* for all materials and compositions. The remarkable fact of nature is that this ratio is not just *constant*—it's exactly equal to 1 for every object ever tested.

Modern experiments have confirmed this equality to extraordinary precision. Roland Eötvös, working in the early 1900s, showed that inertial and gravitational mass are equal to within one part in a billion. Today's experiments have pushed this precision to better than one part in a trillion, making the equivalence of inertial and gravitational mass one of the most precisely verified principles in all of physics.

Einstein's Leap of Insight

What transformed this curious observational fact into a revolutionary principle was Einstein's genius for seeing deeper connections. Instead of accepting the equality of inertial and gravitational mass as a strange coincidence, Einstein asked a more fundamental question: What if they're equal because they're actually the same thing?

This insight led Einstein to what he later called his "happiest thought"—the realization that a person in free fall experiences no gravitational effects at all. Imagine Elena in that elevator with the severed cable. During her terrifying descent, she would float weightlessly in the center of the elevator car. If she released a pen from her hand, it wouldn't fall to the floor—it would simply float beside her, both she and the pen falling together under gravity's influence.

More dramatically, from Elena's perspective inside the falling elevator, gravity would seem to have vanished completely. She could perform any physics experiment she wanted—bouncing balls, swinging pendulums, measuring the trajectories of projectiles—and the results would be identical to what she would observe in a laboratory floating in the vacuum of deep space, far from any gravitational influence.

This led Einstein to a profound realization: the effects of gravity can be completely eliminated locally by choosing an appropriate reference frame—specifically, a freely falling reference frame. This wasn't just a mathematical trick; it revealed something fundamental about the nature of gravity itself.

The Elevator Thought Experiment

Einstein's most famous illustration of the equivalence principle involves imagining yourself in a windowless elevator under two different circumstances. This thought experiment, simple as it seems, contains the seeds of our modern understanding of gravity, spacetime, and time dilation.

Scenario 1: You're standing in an elevator on Earth's surface. You feel your normal weight pressing down on the elevator floor. Drop a ball, and it falls toward the floor with acceleration $g = 9.8$ meters per second squared. From your perspective, gravity pulls the ball downward.

Scenario 2: You're in an identical elevator, but now it's in deep space, far from any massive objects, attached to a rocket that's accelerating upward at exactly 9.8 meters per second squared. You feel exactly the same sensation of weight as your body is pressed against the elevator floor by the rocket's acceleration. Drop the same ball, and it appears to fall toward the floor with the same acceleration, but now it's because the elevator floor is accelerating upward to meet the ball.

Einstein's key insight was that these two scenarios are physically indistinguishable. No experiment you could perform inside the windowless elevator could tell you whether you're experiencing gravitational attraction or rocket-powered acceleration. The equivalence principle states that gravitational fields and accelerated reference frames are locally equivalent—not just similar, but genuinely indistinguishable.

This equivalence extends to all physical phenomena. Light beams, electromagnetic fields, radioactive decay, atomic transitions—every physical process behaves identically in a gravitational field and in an equivalently accelerated reference frame. This isn't just a useful approximation; it's a fundamental principle that reveals the deep structure of spacetime itself.

The Principle of General Covariance

The equivalence principle led Einstein to an even more general insight: the laws of physics should be the same for all observers, regardless of their state of motion—not just for observers in uniform motion, as required by special relativity, but for observers undergoing any kind of acceleration. This principle, known as general covariance, was the foundation upon which Einstein built his general theory of relativity.

If the laws of physics are truly universal, they must be expressible in a form that doesn't depend on the particular coordinate system chosen to describe them. Just as the distance between two points doesn't depend on whether you measure it in feet or meters, the fundamental laws of nature shouldn't depend on whether you describe them from the perspective of someone standing on Earth, someone falling in an elevator, or someone accelerating through space.

This requirement of general covariance placed severe constraints on the possible forms that physical laws could take. Einstein realized that the mathematics needed to describe such generally covariant laws was differential geometry—the same mathematical framework used to describe curved surfaces and spaces. This mathematical connection wasn't accidental; it was pointing toward a profound physical truth.

From Equivalence to Curvature

The equivalence principle led Einstein to perhaps his most revolutionary insight: gravity is not a force in the traditional sense, but rather a manifestation of the curvature of spacetime itself. To understand this profound idea, consider what the equivalence principle tells us about the nature of gravitational motion.

In a freely falling reference frame, gravity disappears—objects move in straight lines at constant velocities, just as they do in special relativity. But what happens when we observe this motion from a different reference frame, one that's not in free fall? From such a frame, the "straight line" motion of free-falling objects appears curved.

Einstein realized that this curvature wasn't due to a force deflecting objects from their natural straight-line paths. Instead, the objects were following the straightest possible paths through a curved spacetime. Just as a ship following a great circle route on Earth's curved surface appears to curve when viewed on a flat map, objects following straight paths through curved spacetime appear to curve when viewed from certain reference frames.

This insight transformed our understanding of gravity. Instead of thinking of massive objects as exerting forces on other objects, we now understand that massive objects curve the spacetime around them, and other objects simply follow the straightest possible paths through this curved spacetime geometry. What we experience as gravitational attraction is actually the result of moving through curved spacetime.

Local Versus Global Effects

The equivalence principle applies locally—within small regions of spacetime where gravitational fields can be approximated as uniform. This local character is crucial because real gravitational fields are not perfectly uniform; they vary from place to place due to tidal effects.

Consider the difference between being in a small elevator falling toward Earth versus being in a large room falling toward Earth. In the small elevator, every part of the space experiences essentially the same gravitational field, so the equivalence principle works perfectly—gravity can be completely eliminated by free fall.

But in the large room, different parts of the space experience slightly different gravitational fields. Objects closer to Earth experience slightly stronger gravity than objects farther away. This creates tidal effects—gradual stretching and compression that reveal the presence of gravity even in free fall.

These tidal effects are manifestations of spacetime curvature. In perfectly flat spacetime, free-falling observers would never observe any relative acceleration between nearby objects. But in curved spacetime, even freely falling observers can detect gradual relative motions that reveal the underlying curvature.

The mathematics of general relativity captures this distinction precisely. The equivalence principle allows us to eliminate the effects of gravity locally by choosing freely falling coordinates, but the curvature of spacetime manifests as tidal effects that cannot be eliminated by any choice of coordinates.

Experimental Tests of Equivalence

The equivalence principle has been tested with extraordinary precision, making it one of the most thoroughly verified principles in physics. These tests fall into several categories, each probing different aspects of the principle's validity.

Eötvös-type experiments test the weak equivalence principle by comparing the gravitational acceleration of different materials. Modern versions of these experiments can detect differences in the ratio of gravitational to inertial mass at the level of parts in 10^{13} or better. Despite this incredible precision, no violations of the equivalence principle have ever been detected.

Lunar laser ranging experiments provide another powerful test. By measuring the precise distance to retroreflectors placed on the Moon's surface by Apollo astronauts and Soviet robotic missions, scientists can test whether the Earth and Moon fall toward the Sun at exactly the same rate, despite their very different compositions. After decades of measurements, these experiments confirm the equivalence principle to extraordinary precision.

Satellite experiments have pushed the precision even further. The MICROSCOPE satellite, launched by the European Space Agency, tested the equivalence principle in the nearly perfect vacuum of space, achieving precision levels better than one part in 10^{15}. Future missions may push this precision to even more extraordinary levels.

The remarkable thing about all these experiments is not just their precision, but their consistency. Despite testing the equivalence principle with materials ranging from ordinary metals to exotic atomic nuclei, from everyday laboratory masses to astronomical bodies, every experiment confirms Einstein's insight that gravitational and inertial mass are not just approximately equal—they are exactly identical.

Implications for Time Dilation

For our exploration of time dilation, the equivalence principle has profound implications. If gravity and acceleration are truly equivalent, then gravitational fields should affect the passage of time in the same way that acceleration does in special relativity. This connection leads directly to the phenomenon of gravitational time dilation.

Consider the elevator thought experiment once more, but now focus on what happens to light and time. If the elevator is accelerating upward through space, a light beam entering horizontally will appear to curve downward from the perspective of someone in the elevator. This curvature occurs because during the time the light crosses the elevator, the elevator has accelerated upward, changing its velocity relative to the light beam.

But the equivalence principle tells us that this accelerating elevator is indistinguishable from an elevator at rest in a gravitational field. Therefore, light must also curve in gravitational fields—an effect that was spectacularly confirmed during the 1919 solar eclipse expedition led by Arthur Eddington.

More subtly, if the elevator is accelerating, then clocks at different heights within the elevator will appear to run at different rates. A clock at the top of the elevator, farther from the direction of acceleration, will appear to run faster than a clock at the bottom. By the equivalence principle, this means that clocks at different gravitational potentials—different heights in a gravitational field—should also run at different rates.

This is gravitational time dilation, and it represents one of the most remarkable consequences of the equivalence principle. Unlike the velocity-based time dilation we explored in earlier chapters, gravitational time dilation doesn't require any motion through space—it arises purely from the presence of gravitational fields, or equivalently, from the curvature of spacetime itself.

The Redshift Prediction

One of Einstein's earliest applications of the equivalence principle was to predict the gravitational redshift—the change in frequency of light as it moves through gravitational fields. This prediction provided one of the first experimental tests of general relativity and demonstrated the intimate connection between gravity and time.

Consider light traveling upward in a gravitational field, from a region of stronger gravity to a region of weaker gravity. From the perspective of the equivalence principle, this is equivalent to light traveling from the bottom of an accelerating elevator to the top. During the time the light takes to make this journey, the elevator accelerates, changing its velocity relative to the light source.

This relative motion creates a Doppler shift—the light appears redshifted (lower frequency) when observed at the top of the elevator. By the equivalence principle, the same effect must occur in gravitational fields: light traveling upward against gravity must be redshifted, while light traveling downward with gravity must be blueshifted.

But frequency is intimately related to time. If light from a source deep in a gravitational field appears redshifted when observed at a higher altitude, this suggests that time itself passes more slowly in stronger gravitational fields. Clocks in stronger gravity literally tick more slowly than identical clocks in weaker gravity.

This gravitational time dilation effect has been confirmed with extraordinary precision using atomic clocks. Even the tiny difference in gravitational potential between the ground floor and the top floor of a building is sufficient to create measurable time dilation effects with modern atomic clocks.

The Geometry of Spacetime

Perhaps the most profound implication of the equivalence principle is its revelation that spacetime itself has geometry—and that this geometry is curved by the presence of matter and energy. This geometric interpretation of gravity represents one of the most beautiful unifications in the history of physics.

In Newton's universe, space and time provided a fixed stage on which physical events unfolded. Gravity was a force that acted between objects across this unchanging backdrop. But Einstein's equivalence principle revealed that space and time are not passive observers but active participants in physical phenomena.

Massive objects curve the spacetime around them, and this curvature is what we experience as gravity. Objects in free fall are not being acted upon by a gravitational force—they are simply following the straightest possible paths (geodesics) through curved spacetime. The curvature guides their motion in exactly the way that Newton's gravitational force was thought to do.

This geometric interpretation explains why the equivalence principle works so perfectly. In a small enough region of curved spacetime, the curvature appears approximately flat, just as a small region of Earth's curved surface appears approximately flat to someone standing on it. In such regions, the laws of special relativity apply, and the effects of curvature (gravity) can be eliminated by choosing freely falling coordinates.

The Gateway to General Relativity

The equivalence principle served as Einstein's gateway to the full theory of general relativity. Once he realized that gravity and acceleration were equivalent, and that this equivalence revealed the geometric nature of spacetime, Einstein needed to develop the mathematical framework to describe curved spacetime quantitatively.

This led him to the mathematics of differential geometry, particularly the work of Bernhard Riemann on curved spaces. With the help of his friend Marcel Grossmann, Einstein learned to use tensors, metrics, and other geometric tools to describe the curvature of spacetime.

The culmination was Einstein's field equations, published in 1915, which describe precisely how matter and energy curve spacetime, and how this curvature guides the motion of objects and the propagation of light. These equations encode the equivalence principle at their very heart—they ensure that in any local region of spacetime, an appropriate choice of coordinates can eliminate the effects of gravity and reduce the physics to that of special relativity.

Modern Applications and Extensions

Today, the equivalence principle continues to play a crucial role in both theoretical physics and practical applications. Every GPS satellite relies on corrections based on the equivalence principle to maintain timing accuracy. Every test of general relativity fundamentally depends on the validity of equivalence.

Modern theoretical physics continues to explore the implications and limitations of the equivalence principle. In quantum gravity theories, physicists investigate whether the principle remains valid at the smallest scales of space and time. String theory and other approaches to unifying gravity with quantum mechanics must grapple with how the equivalence principle emerges from more fundamental principles.

The search for violations of the equivalence principle continues with ever-increasing precision. While no violations have been detected, any deviation from perfect equivalence would provide crucial clues about the nature of gravity and its relationship to other fundamental forces. Future space-based experiments may push the precision of equivalence tests to levels that could reveal new physics beyond Einstein's theory.

Philosophical Revolution

Beyond its technical implications, the equivalence principle represents a profound philosophical revolution in our understanding of the relationship between geometry and physics.

Before Einstein, geometry was considered a branch of mathematics, completely separate from physical science. Space was thought to have a fixed Euclidean geometry that provided the unchanging stage for physical events.

The equivalence principle revealed that geometry and physics are intimately connected. The geometry of spacetime is not fixed but dynamic, shaped by the matter and energy it contains. This geometric interpretation of gravity showed that mathematics—specifically, the mathematics of curved spaces—could describe the deepest aspects of physical reality.

This insight has had profound implications extending far beyond gravity. Modern physics routinely uses geometric concepts to understand electromagnetic fields, quantum mechanics, and the other fundamental forces. The unification of geometry and physics that began with the equivalence principle continues to guide our understanding of the universe's deepest structure.

Preparing for Gravitational Time Dilation

As we've seen throughout this chapter, the equivalence principle opens the door to understanding one of the most remarkable phenomena in Einstein's universe: gravitational time dilation. The principle tells us that clocks in gravitational fields must behave exactly as they do in accelerated reference frames, leading to the conclusion that time itself runs slower in stronger gravitational fields.

This gravitational effect on time can be far more dramatic than the velocity-based time dilation we explored in earlier chapters. While achieving significant velocity-based time dilation requires speeds approaching the cosmic speed limit, significant gravitational time dilation can occur in the presence of sufficiently strong gravitational fields—fields that exist around neutron stars, black holes, and other exotic astronomical objects.

The equivalence principle provides the conceptual foundation for understanding these extreme effects. By establishing that gravity and acceleration are equivalent, and that both affect the passage of time, Einstein opened the door to understanding how the curvature of spacetime itself can stretch and compress time in ways that seem almost magical but are precisely calculable using the mathematics of general relativity.

The Elegance of Equivalence

What makes the equivalence principle so remarkable is its combination of simplicity and profundity. The basic insight—that you can't tell the difference between gravity and acceleration—can be understood by anyone who has felt weightless in a falling elevator or experienced artificial gravity in a spinning spacecraft. Yet this simple observation led to a complete revolution in our understanding of space, time, and gravity.

The principle embodies Einstein's genius for finding deep connections where others saw only coincidence. Instead of accepting the equality of inertial and gravitational mass as a curious accident of nature, Einstein recognized it as a clue to something much more fundamental about the structure of reality itself.

This transformation of an observational curiosity into a fundamental principle illustrates how the greatest advances in physics often come not from discovering new phenomena but from seeing familiar phenomena in a completely new light. The equivalence principle took facts that had been known for centuries—that all objects fall at the same rate, that gravity and acceleration feel similar—and revealed their profound implications for the nature of spacetime itself.

As we prepare to explore the specific mechanisms of gravitational time dilation in the next chapter, we carry with us Einstein's profound insight that gravity is not a force acting through space and time but rather a manifestation of the curvature of spacetime itself. This curvature affects not just the motion of objects but the very passage of time, creating effects that can be far more dramatic than anything we encounter in our everyday experience of Einstein's remarkable universe.

The equivalence principle stands as one of the most elegant and far-reaching insights in the history of human thought—a simple observation about falling elevators that unlocked the secrets of space, time, and gravity themselves. In revealing that gravity and acceleration are equivalent, Einstein showed us that the universe is even stranger and more beautiful than we had dared to imagine, while simultaneously providing the tools to understand and calculate the effects of this strangeness with mathematical precision.

Chapter 8: Spacetime - The Fabric of Reality

Professor Michael Harrison stood at the edge of the Large Hadron Collider's visitor center, gazing down at a massive steel tube that stretched beyond the horizon in both directions. Twenty-seven kilometers of the world's most sophisticated machinery lay buried 100 meters beneath his feet, designed to smash protons together at energies so extreme that they recreate conditions from the first moments after the Big Bang. But what fascinated Harrison most as a theoretical physicist wasn't the impressive engineering—it was the invisible arena in which these cosmic-energy collisions unfold.

Every proton racing through those superconducting magnets, every collision that creates exotic particles lasting mere trillionths of a second, every detector measurement that confirms or challenges our theories—all of it takes place within the fabric of spacetime itself. This fabric isn't just the passive stage on which physics unfolds; it's an active participant, bending and flexing in response to the matter and energy it contains, guiding the very phenomena that the LHC was built to study.

Standing there in the Swiss countryside, Harrison was contemplating one of the most profound insights in human history: that space and time aren't separate entities providing a fixed backdrop for physical events, but rather a unified, dynamic fabric that can be stretched, compressed, twisted, and curved by the matter and energy within it. This fabric—spacetime—is the true arena of reality, and understanding its properties is the key to comprehending how gravity can stretch and compress time itself.

The Marriage of Space and Time

Before Einstein, space and time seemed as separate and distinct as any two concepts could be. Space was the three-dimensional arena in which objects moved and events occurred—up and down, left and right, forward and backward. Time was the universal clock that ticked steadily for everyone, everywhere, measuring the duration between events and providing the temporal framework against which all change could be measured.

Newton's physics reinforced this separation by treating space as absolute—a fixed, unchanging container that existed independently of the matter it held—and time as absolute—flowing at the same steady rate throughout the entire universe regardless of what was happening within it. This seemed not just reasonable but obviously correct based on everyday experience.

But Einstein's special theory of relativity shattered this comfortable separation forever. His insights revealed that space and time are not independent entities but rather different aspects of a single, unified reality. Events that appear simultaneous to one observer occur at different times for another observer in relative motion. Distances that seem fixed to one observer appear contracted to another. The duration between events depends on the motion of whoever measures it.

These aren't just measurement effects or observational artifacts—they reflect the fundamental structure of reality itself. Space and time are woven together so intimately that it makes no sense to speak of them separately. Einstein showed us that we live in a four-dimensional spacetime, where the three familiar dimensions of space (length, width, and height) combine with one dimension of time to create the unified fabric of reality.

This unification was more than just a mathematical convenience. It revealed that space and time respond to matter and motion in coordinated ways that preserve deeper symmetries. When objects move through space at high speeds, both space and time adjust—distances contract and time dilates—in precisely the amounts needed to keep the speed of light constant for all observers.

The Dynamic Fabric

Special relativity revealed that spacetime exists, but it took Einstein's general theory of relativity to show us that spacetime is dynamic—that it can bend, flex, and curve in response to the matter and energy within it. This insight transformed our understanding of gravity from a mysterious force acting across empty space to a manifestation of spacetime geometry itself.

The key realization came from the equivalence principle we explored in the previous chapter. If gravity and acceleration are truly equivalent, and if acceleration affects the geometry of spacetime (as special relativity showed), then gravity must also be geometric in nature. Einstein realized that what we experience as gravitational attraction is actually the result of following straight paths through curved spacetime.

Imagine trying to walk in a straight line across a gently rolling hillside. Even though you're walking straight from your own perspective, someone observing from above would see your path curve as you follow the contours of the landscape. Similarly, objects moving through curved spacetime follow paths that appear curved to distant observers, even though the objects are moving along the straightest possible routes through the geometry they inhabit.

This geometric interpretation of gravity was revolutionary because it eliminated the mysterious "action at a distance" that had troubled scientists since Newton's time. In Newton's theory, massive objects somehow reached across empty space to pull on other objects with gravitational force. But Einstein showed that there's no mysterious action at a distance—massive objects simply curve the spacetime around them, and other objects follow the geometry of that curved spacetime.

Einstein's Masterpiece: The Field Equations

The mathematical heart of general relativity lies in Einstein's field equations, published in their final form in November 1915. These equations capture one of the most elegant relationships in all of physics, often summarized in the phrase: "matter tells spacetime how to curve, and curved spacetime tells matter how to move".

The field equations can be written compactly as $G\mu\nu=8\pi T\mu\nu$, but this mathematical notation conceals the profound physics within. The left side of the equation, $G\mu\nu$, describes the curvature of spacetime—how much the geometry deviates from flat, Euclidean space. The right side, $T\mu\nu$, describes the matter and energy content—the sources that create curvature.

What makes these equations so remarkable is their simplicity and generality. They don't specify what kind of matter creates curvature—whether it's ordinary atoms, exotic particles, electromagnetic fields, or even other gravitational fields. They simply state that any form of matter or energy will curve spacetime in proportion to its density, and that this curvature will guide the motion of everything in the vicinity.

The equations also reveal that spacetime curvature affects itself. Unlike Newton's gravity, where gravitational fields don't interact with each other, Einstein's equations show that gravitational fields carry energy and can therefore create additional curvature. This nonlinearity makes the equations notoriously difficult to solve, but it also leads to phenomena like gravitational waves—ripples in spacetime itself that propagate at the speed of light.

The Rubber Sheet Universe

Perhaps no analogy has been more useful for understanding curved spacetime than the image of a rubber sheet stretched taut and then deformed by placing massive objects on it. Imagine a perfectly flat, elastic membrane representing spacetime. When you place a bowling ball (representing a massive object like the Sun) in the center, it creates a depression in the sheet. Now, if you roll a marble (representing a planet) across the sheet, it will follow a curved path around the depression, potentially entering a stable orbit around the bowling ball.

This behavior occurs not because the marble is "attracted" to the bowling ball by some mysterious force, but because it's following the straightest possible path across a surface that has been curved by the bowling ball's presence. The marble is simply responding to the geometry of the space it inhabits.

While this analogy is incredibly useful, it's important to recognize its limitations. The rubber sheet represents only two dimensions of space plus the curvature, while real spacetime has three dimensions of space plus one of time, all of which can be curved. Also, the analogy relies on Earth's gravity to make the bowling ball sink into the sheet, which can create confusion since we're trying to explain gravity itself.

Despite these limitations, the rubber sheet analogy captures the essential insight: massive objects don't pull on other objects through some mysterious force; instead, they curve the spacetime around them, and objects simply follow the geometry of that curved spacetime. What we experience as gravitational attraction emerges naturally from the geometry of curved spacetime.

The Geometry of Gravity

To understand how spacetime curvature creates what we experience as gravity, we need to think about what "straight lines" mean in curved space. In flat, Euclidean geometry, the shortest distance between two points is a straight line. But on curved surfaces, the shortest paths—called geodesics—appear curved when viewed from certain perspectives.

Consider the surface of Earth. The shortest flight path between New York and London doesn't appear straight on a typical flat map—it curves northward over Greenland and the North Atlantic. But this great circle route is actually the straightest possible path across Earth's curved surface. Pilots follow these curved routes not because they're being deflected by some force, but because they're taking the most direct path through the geometry they inhabit.

Spacetime curvature works similarly, but with the added complexity of involving time as well as space. When objects move through curved spacetime, they follow geodesics—the straightest possible paths through four-dimensional spacetime geometry.

These geodesics appear curved in space to distant observers, creating what we interpret as gravitational acceleration.

The mathematics of general relativity provides precise tools for calculating these geodesics. Given the distribution of matter and energy in a region of spacetime, Einstein's field equations determine the curvature, and the geodesic equations then specify exactly how objects will move through that curved geometry. This mathematical framework has proven extraordinarily successful at predicting everything from planetary orbits to the behavior of light near black holes.

Time as Part of the Fabric

One of the most profound aspects of spacetime curvature is that it affects not just the spatial dimensions but also the temporal dimension—time itself becomes part of the curved geometry. This is what creates gravitational time dilation, one of the most remarkable consequences of Einstein's theory.

When spacetime is curved by the presence of mass and energy, the geometry affects how time flows at different locations. Clocks closer to massive objects—deeper in gravitational "wells"—tick more slowly than identical clocks farther away. This isn't a mechanical effect on the clocks themselves; it's a fundamental property of how time works in curved spacetime.

The connection between spacetime curvature and time dilation can be understood through the rubber sheet analogy, though we must extend it to include the temporal dimension. Just as the spatial geometry becomes curved near massive objects, the temporal geometry also becomes distorted. Time "flows" more slowly where spacetime curvature is more extreme.

This temporal curvature has been measured with extraordinary precision using atomic clocks at different altitudes. Even the tiny difference in gravitational potential between the ground floor and upper floors of tall buildings creates measurable time dilation effects. GPS satellites must constantly correct for the fact that their clocks run faster than Earth-based clocks due to the weaker gravitational field at orbital altitude.

The Dynamic Nature of Spacetime

Unlike Newton's absolute space and time, which provided a fixed, unchanging backdrop for physical events, Einstein's spacetime is dynamic—it responds to the matter and energy within it and can change over time. This dynamic character leads to some of the most remarkable predictions of general relativity.

Spacetime can vibrate, creating gravitational waves that propagate at the speed of light. These ripples in the fabric of reality itself were predicted by Einstein but weren't directly detected until 2015, when the LIGO experiment observed gravitational waves from the collision of two black holes billions of years ago.

Spacetime can also expand or contract on cosmic scales. Einstein's field equations, when applied to the universe as a whole, predict that spacetime itself is either expanding or contracting. This prediction, initially so counterintuitive that Einstein tried to modify his equations to avoid it, was confirmed by Edwin Hubble's observations of distant galaxies and is now understood as the expansion of the universe itself.

The dynamic nature of spacetime also means that extreme concentrations of matter and energy can create exotic geometries—regions where spacetime is curved so severely that it creates phenomena like black holes, where the curvature becomes so extreme that not even light can escape.

Frame Dragging and Spacetime Dynamics

One of the most subtle predictions of general relativity involves frame dragging—the tendency for rotating masses to "drag" spacetime along with them as they spin. This effect demonstrates that spacetime doesn't just respond to the presence of mass and energy but also to their motion.

When a massive object rotates, it creates a swirling pattern in the spacetime around it, similar to how a spoon stirring honey creates swirling currents in the liquid. Objects near a rotating mass experience this frame dragging as a tendency to be carried along with the rotation, even if they're not in direct contact with the rotating object.

Frame dragging has been measured around Earth using satellite experiments that detect tiny changes in the orientation of gyroscopes in orbit. These measurements confirm that Earth's rotation does indeed drag spacetime along with it, exactly as Einstein's equations predict. The effect is small around Earth, but it becomes dramatic near rapidly rotating black holes, where the frame dragging can approach the speed of light.

The Hierarchy of Curvature Effects

Spacetime curvature manifests differently at different scales and in different environments, creating a rich hierarchy of gravitational phenomena. Near Earth's surface, spacetime is only very weakly curved—the deviations from flat geometry are tiny, which is why Newton's approximation works so well for everyday purposes.

But as we move to more extreme environments, curvature effects become more pronounced. Near the Sun, spacetime curvature is strong enough to deflect light from distant stars by measurable amounts—the effect that made Einstein famous when it was observed during the 1919 solar eclipse. Near white dwarf stars, curvature effects become strong enough to cause significant gravitational time dilation and redshift of light.

Around neutron stars—objects so dense that a teaspoon of their material would weigh as much as Mount Everest—spacetime curvature becomes extreme. Time dilation effects are dramatic, light is bent through large angles, and the very atoms of matter are crushed into exotic states that can't exist anywhere else in the universe.

At the edge of black holes, spacetime curvature reaches its ultimate extreme. The curvature becomes so severe that it creates an event horizon—a boundary beyond which nothing, not even light, can escape. Inside the event horizon, the roles of space and time become interchanged in ways that challenge our most basic intuitions about the nature of reality.

Experimental Confirmations

The predictions of spacetime curvature have been confirmed by countless experiments and observations, making general relativity one of the most thoroughly tested theories in physics. The bending of light near the Sun, first observed during solar eclipses, has been measured with increasing precision using radio astronomy and spacecraft tracking.

The precession of Mercury's orbit—the slow rotation of the planet's elliptical path around the Sun—exactly matches Einstein's predictions for motion through curved spacetime.

Binary pulsar systems, where neutron stars orbit each other at tremendous speeds, provide natural laboratories for testing relativity in extreme gravitational fields.

Perhaps most dramatically, the recent detection of gravitational waves by LIGO and other interferometric detectors has provided direct evidence that spacetime itself can ripple and vibrate, carrying energy across the universe at the speed of light. These observations confirm that spacetime is indeed the dynamic fabric that Einstein envisioned.

The Quantum Challenge

While Einstein's description of spacetime as a smooth, continuous fabric has been extraordinarily successful, modern physics suggests that this picture may be incomplete. Quantum mechanics implies that at the smallest scales—distances comparable to the Planck length, about 10^{-35} meters—spacetime itself may become quantum mechanical and discontinuous.

At these unimaginably tiny scales, the smooth geometry of Einstein's spacetime may give way to a foam-like structure where space and time themselves fluctuate quantum mechanically. Understanding how to reconcile Einstein's geometric view of spacetime with quantum mechanics remains one of the greatest challenges in theoretical physics.

Various approaches to quantum gravity—including string theory, loop quantum gravity, and other frameworks—attempt to describe how spacetime behaves when quantum effects become important. While these theories remain speculative, they all must account for the success of Einstein's geometric description at macroscopic scales while extending it to the quantum realm.

Spacetime and Consciousness

The recognition that we live in curved spacetime has profound implications that extend beyond physics. If spacetime is the arena in which all physical processes unfold, then consciousness itself—whatever its ultimate nature—must emerge from processes occurring within this curved geometric framework.

This raises deep questions about the relationship between subjective experience and objective spacetime geometry. Our perception of "now"—the seemingly universal present moment—has no counterpart in Einstein's spacetime, where simultaneity is relative and depends on motion through space.

Yet somehow, conscious experience emerges from physical processes unfolding in this four-dimensional curved spacetime. The flow of subjective time, the experience of causation, and our sense of spatial orientation all arise from neural processes governed by the geometry of spacetime. Understanding this connection between geometry and consciousness remains one of the deepest unsolved problems in science.

The Cosmological Implications

Einstein's insight that spacetime is dynamic has revolutionized our understanding of the universe's origin, evolution, and ultimate fate. The field equations, when applied to the universe as a whole, describe how spacetime itself expands, carrying galaxies along with it like raisins in rising bread dough.

This expanding spacetime provides a natural explanation for why distant galaxies appear to be receding from us—they're not moving through space but rather being carried along by the expansion of space itself. The cosmic microwave background radiation, the afterglow of the Big Bang, provides a snapshot of the universe when it was young and hot, confirming many details of how spacetime has evolved over cosmic time.

The geometry of spacetime on cosmic scales determines the universe's ultimate fate. If spacetime contains enough matter and energy, its expansion will eventually halt and reverse, leading to a "Big Crunch." If it contains too little, expansion will continue forever. The precise balance depends on the total matter and energy content of the universe, including mysterious components like dark matter and dark energy.

The Beauty of Geometric Unity

What makes Einstein's spacetime so remarkable is how it unifies phenomena that previously seemed completely unrelated. Gravity, which Newton described as a force acting across empty space, becomes geometry. Time dilation, which seemed like an exotic consequence of high-speed motion, becomes a natural feature of curved spacetime. The precession of planetary orbits, the bending of light, the expansion of the universe—all emerge from the single insight that spacetime is curved by matter and energy.

This geometric unity reflects one of the deepest principles in physics: the idea that nature operates according to mathematical principles of extraordinary beauty and elegance. The field equations that describe spacetime curvature possess a mathematical beauty that transcends their practical applications, suggesting deep connections between geometry, physics, and the structure of reality itself.

Einstein once remarked that "the most incomprehensible thing about the universe is that it is comprehensible." The geometric nature of spacetime exemplifies this profound truth—the fact that the arena of reality itself can be described by mathematical equations reveals something remarkable about the relationship between human reason and the structure of the cosmos.

Preparing for Gravitational Time Dilation

As we've explored throughout this chapter, spacetime is not the passive backdrop that Newton envisioned but an active participant in physical phenomena—a dynamic fabric that can be stretched, compressed, and curved by the matter and energy within it. This curvature affects not just the motion of objects through space but also the flow of time itself.

Understanding spacetime as a curved, dynamic fabric provides the foundation for comprehending gravitational time dilation—the phenomenon where time itself runs slower in stronger gravitational fields. Unlike the velocity-based time dilation we explored earlier, which requires motion through space at significant fractions of light speed, gravitational time dilation arises purely from the curvature of spacetime itself.

The effects can be far more dramatic than those produced by motion alone. While achieving extreme velocity-based time dilation requires enormous energies to approach light speed, extreme gravitational time dilation occurs naturally around neutron stars, black holes, and other exotic objects where spacetime curvature becomes severe.

The geometric perspective of curved spacetime provides both the conceptual framework and the mathematical tools for understanding these effects. Einstein's field equations tell us how matter and energy curve spacetime, while the geodesic equations describe how objects—including light rays and time itself—respond to that curvature.

The Fabric of Everything

In the end, spacetime is the fabric from which everything in our universe is woven. Every particle, every force, every thought, and every moment of consciousness unfolds within this four-dimensional geometric arena. The curvature of spacetime guides the formation of galaxies and the evolution of stars, determines the orbits of planets and the paths of light rays, and shapes the very flow of time itself.

Professor Harrison, standing above the LHC, was contemplating this profound truth: that the exotic particles created in high-energy collisions, the detectors that record their brief existence, and even his own consciousness as he pondered their significance—all of it exists within and is shaped by the curved geometry of spacetime.

This fabric isn't just the stage on which the cosmic drama unfolds; it's an active participant in every scene. From the quantum fluctuations that may have seeded the formation of galaxies to the gravitational waves that ripple across the cosmos, spacetime itself is constantly evolving, responding to the matter and energy within it while simultaneously guiding their behavior through its geometric properties.

As we prepare to explore how this curved spacetime creates gravitational time dilation effects that can slow time to a crawl or stretch it beyond recognition, we carry with us Einstein's profound insight: that space and time are not separate entities but aspects of a unified fabric that shapes—and is shaped by—everything within it. The universe is not just stranger than we imagine; it's more beautiful, more unified, and more comprehensible than anyone dared to hope.

The fabric of spacetime is the canvas on which the story of time dilation is written, and as we'll discover in the coming chapters, that story involves some of the most extreme and remarkable phenomena in the universe—places where Einstein's insights reveal time behaving in ways that challenge our deepest intuitions while confirming the mathematical beauty of the laws that govern our remarkable cosmos.

Chapter 9: Clocks in the Gravity Well

Dr. Sarah Chen stood on the observation deck of the Tokyo Skytree, 634 meters above the bustling streets of Japan's capital, holding two identical atomic clocks that had been synchronized that morning in her laboratory at sea level. As she looked down at the sprawling city below, she was about to witness one of the most profound demonstrations of Einstein's theory in action—a demonstration that would reveal how gravity itself reaches into the very mechanism of time and subtly, inexorably, alters its flow.

The atomic clock in her hands was ticking at a rate that was imperceptibly, but measurably, faster than an identical clock sitting on her laboratory bench 634 meters below. Not because of any mechanical difference between the devices, not because of temperature variations or electromagnetic interference, but because time itself—the fundamental fabric of temporal reality—was flowing at a different rate at the top of the Tokyo Skytree than it was at ground level.

This phenomenon, known as gravitational time dilation, represents one of the most remarkable and counterintuitive consequences of Einstein's general theory of relativity. Unlike the velocity-based time dilation we explored in earlier chapters, which requires motion through space at significant fractions of light speed, gravitational time dilation arises from something far more subtle: the simple act of being at different positions within a gravitational field.

What Dr. Chen was experiencing atop the Tokyo Skytree demonstrates that we all live our lives sliding up and down through what physicists call "gravity wells"—regions of curved spacetime where the strength of gravitational fields varies from place to place. Every time we climb a flight of stairs, ride an elevator, or even stand up from a chair, we're moving to regions where time itself flows at a slightly different rate.

The Principle Behind the Effect

Gravitational time dilation emerges naturally from the equivalence principle we explored in Chapter 7—the insight that gravitational fields and accelerated reference frames are locally indistinguishable. If acceleration affects the passage of time, as special relativity demonstrates, then gravity must affect time in precisely the same way.

To understand this connection, imagine yourself in an elevator accelerating upward through space at 9.8 meters per second squared—exactly the same acceleration you experience due to Earth's gravity when standing on the surface. From the equivalence principle, the physics inside this accelerating elevator should be identical to the physics you experience in Earth's gravitational field.

Now consider what happens to light and time in the accelerating elevator. A photon entering the elevator horizontally will appear to curve downward from your perspective, because during the time it takes the light to cross the elevator, the elevator has accelerated upward, changing its velocity relative to the light beam.

More subtly, if there are clocks at different heights within the elevator, they will appear to run at different rates. A clock at the ceiling of the elevator, farther from the direction of acceleration, will appear to tick faster than a clock at the floor.

This occurs because the acceleration creates a gradient in the flow of time throughout the elevator.

By the equivalence principle, these same effects must occur in gravitational fields. Light must bend when passing through gravity (an effect spectacularly confirmed during the 1919 solar eclipse), and clocks at different gravitational potentials—different depths in the gravity well—must tick at different rates.

The Mathematics of Gravitational Time Dilation

The mathematical relationship governing gravitational time dilation is elegantly simple, at least for weak gravitational fields like those we encounter near Earth. If we have two clocks separated by a height difference h in a uniform gravitational field g, the clock at the higher altitude will run faster than the lower clock by a fractional amount:

$\Delta tt = gh/c^2$

Where Δt is the time difference that accumulates, t is the elapsed time, g is the gravitational acceleration, h is the height difference, and c is the speed of light.

Let's apply this formula to Dr. Chen's situation atop the Tokyo Skytree. With $g = 9.8\ m/s^2$, $h = 634\ m$, and $c = 3\times10^8\ m/s$, we get:

$\Delta tt = (9.8\times634)/(3\times10^8)^2 = 6{,}213 \div 9\times10^{16} = 6.9\times10^{-14}$

This means that for every second that passes, the clock at the top of Tokyo Skytree gains about 69 femtoseconds (69 quadrillionths of a second) relative to a clock at ground level. Over the course of a full day (86,400 seconds), the elevated clock would gain about 6 nanoseconds.

While this difference is unimaginably small by human standards, it's easily measurable with modern atomic clocks, which can detect time differences of femtoseconds or even smaller intervals. More importantly, this tiny effect demonstrates that gravitational time dilation is not just a theoretical curiosity but a real, measurable phenomenon that affects every moment of our lives.

The Tallest Buildings and Time

The effect Dr. Chen observed from Tokyo Skytree becomes more pronounced as we consider taller structures. At the top of the Burj Khalifa in Dubai, standing 828 meters above sea level, time runs about 108 femtoseconds per second faster than at ground level. Over a year, a clock at the top of the Burj Khalifa would gain about 3.4 microseconds relative to one at its base.

Even more dramatic effects occur at the tops of mountains. At the summit of Mount Everest, 8,849 meters above sea level, gravitational time dilation causes clocks to run about 970 femtoseconds per second faster than at sea level. A climber who spent a full year at Everest's summit would age about 30 microseconds more than someone who remained at sea level—literally experiencing more time due to the weaker gravitational field at high altitude.

These effects have been directly measured using portable atomic clocks. In one famous experiment, physicists carried atomic clocks to different altitudes and confirmed that the time dilation matched Einstein's predictions with extraordinary precision. The experiments showed that even altitude differences as small as one meter create measurable time dilation effects with sufficiently sensitive instruments.

The GPS Constellation: A Daily Demonstration

Perhaps no technology provides a more dramatic demonstration of gravitational time dilation than the Global Positioning System. GPS satellites orbit Earth at an altitude of approximately 20,200 kilometers, where Earth's gravitational field is significantly weaker than at the surface.

At orbital altitude, each GPS satellite experiences gravitational time dilation that causes its onboard atomic clocks to run about 45 microseconds per day faster than identical clocks on Earth's surface. This effect combines with the velocity-based time dilation from the satellites' orbital motion (which causes the clocks to run about 7 microseconds per day slower) for a net effect of about 38 microseconds per day of faster time.

Without corrections for this gravitational time dilation, GPS would be useless for navigation. Since GPS determines position by measuring the time it takes signals to travel from satellites to receivers, timing errors directly translate into position errors. Light travels about 30 centimeters in one nanosecond, so the 38 microseconds per day of uncorrected time dilation would cause GPS positions to drift by more than 11 kilometers per day.

Every GPS receiver on Earth—from the navigation system in your car to the location services in your smartphone—automatically applies corrections based on Einstein's gravitational time dilation formula. The satellites' onboard computers continuously adjust their clocks to account for the faster passage of time at orbital altitude, while ground-based monitoring stations track and refine these corrections.

This represents one of the most successful large-scale applications of Einstein's theory of relativity. Billions of people rely daily on technology that would be impossible without accounting for gravitational time dilation—a effect that emerges from the curvature of spacetime itself.

Precision Measurements on Earth

The development of increasingly precise atomic clocks has allowed scientists to measure gravitational time dilation over smaller and smaller distance scales. In 2010, researchers at the National Institute of Standards and Technology (NIST) demonstrated that they could measure time dilation effects for height differences of just 33 centimeters—about the height of a typical desk.

Even more remarkably, in 2022, scientists achieved measurements of gravitational time dilation over distances as small as one millimeter. Using an optical atomic clock based on strontium atoms, researchers showed that they could detect the tiny time dilation effect that occurs between the top and bottom of a sample just one millimeter tall.

These extraordinary measurements demonstrate that gravitational time dilation is not just an exotic effect that occurs around neutron stars and black holes—it's a fundamental feature of spacetime that manifests even over the smallest measurable distances. Every step you take up a staircase, every time you stand up from sitting, you're moving between regions where time flows at subtly different rates.

The Pound-Rebka Experiment

One of the most elegant confirmations of gravitational time dilation came from the Pound-Rebka experiment, conducted at Harvard University in 1959. Robert Pound and Glen Rebka used the Mössbauer effect—the recoilless emission and absorption of gamma rays by atomic nuclei—to measure the gravitational redshift of light traveling up through Earth's gravitational field.

The experiment involved gamma ray sources and detectors separated by a vertical distance of just 22.5 meters in the physics building at Harvard. According to Einstein's theory, gamma rays traveling upward against Earth's gravity should lose energy and appear redshifted (lower frequency) when detected at the higher altitude. Conversely, gamma rays traveling downward with gravity should gain energy and appear blueshifted (higher frequency).

The frequency shift predicted by Einstein's theory for this 22.5-meter height difference was incredibly small—about 2.5 parts in 10^{15}. Yet Pound and Rebka were able to measure this tiny effect with remarkable precision, confirming Einstein's predictions to within about 1 percent accuracy.

The gravitational redshift measured in the Pound-Rebka experiment is intimately connected to gravitational time dilation. If light loses energy when climbing out of a gravitational field, this suggests that the source of the light—which could be thought of as a "light clock" emitting regular pulses—is running more slowly in the stronger gravitational field below.

Extreme Gravitational Time Dilation

While the effects we've discussed so far are subtle, gravitational time dilation can become extreme in the vicinity of very massive, compact objects. Around a white dwarf star—a stellar remnant about the size of Earth but containing the mass of the Sun—gravitational time dilation becomes significant enough to observe astronomically.

At the surface of a typical white dwarf, gravitational time dilation causes clocks to run about 50 parts per million slower than they would in free space. This means that for every million seconds (about 11.6 days) that pass for an observer far from the white dwarf, only 999,950 seconds pass at the white dwarf's surface—a difference of nearly one minute.

This effect has been observed in the spectra of light emitted by white dwarf stars. The atomic transitions that produce specific wavelengths of light are like extremely precise clocks, and when these "clocks" run slower due to gravitational time dilation, the light they emit is gravitationally redshifted to longer wavelengths. Astronomical observations of white dwarf spectra show exactly this redshift, confirming Einstein's predictions in the extreme gravitational environments around these compact stellar remnants.

Neutron Stars: Time Dilation Taken to Extremes

The effects become even more dramatic around neutron stars—objects so dense that a teaspoon of their material would weigh as much as Mount Everest. The surface gravity on a neutron star is typically about 100 billion times stronger than Earth's gravity, creating gravitational time dilation effects that challenge our imagination.

At the surface of a typical neutron star, gravitational time dilation causes time to run about 30 percent slower than it does in free space. This means that for every 10 seconds that pass for a distant observer, only about 7 seconds pass at the neutron star's surface. The effect is so strong that it's easily observable in the radiation emitted by these objects.

Neutron stars often emit regular pulses of radio waves, earning them the name "pulsars." These pulses are like cosmic lighthouses, sweeping across space as the neutron star rotates. The pulse timing is affected by gravitational time dilation—the pulses are emitted at a rate determined by the flow of time at the neutron star's surface, which is significantly slower than the flow of time in the nearly empty space through which the pulses travel to reach Earth.

Astronomers can use precise timing measurements of pulsar signals to study the gravitational fields around neutron stars, providing natural laboratories for testing Einstein's theory in extreme conditions that cannot be replicated on Earth. These observations consistently confirm the predictions of general relativity, even in gravitational fields billions of times stronger than anything we experience on our planet.

Black Holes: The Ultimate Gravity Wells

The most extreme gravitational time dilation effects occur near black holes—regions of spacetime where gravity becomes so strong that not even light can escape. At the event horizon of a black hole, gravitational time dilation becomes infinite from the perspective of distant observers.

Consider what happens to an astronaut falling toward a black hole. As she approaches the event horizon, gravitational time dilation causes her clocks to run slower and slower relative to observers watching from a safe distance. To these distant observers, the astronaut's fall appears to slow down dramatically as she approaches the event horizon. Her image becomes increasingly redshifted and dim, but she never quite seems to cross the boundary.

From the astronaut's own perspective, however, her clocks continue to tick normally. She experiences no unusual passage of time as she crosses the event horizon and falls toward the black hole's center. The extreme time dilation is only apparent when comparing her time to that of observers far from the black hole.

This dramatic difference in time flow near black holes has been portrayed in science fiction films, but the reality is even more remarkable than fiction. The gravitational time dilation near black holes represents the ultimate expression of Einstein's insight that space and time form a unified fabric that can be stretched and compressed by the presence of mass and energy.

Time Dilation and the GPS Clock Network

The practical importance of gravitational time dilation extends beyond individual GPS satellites to the entire network of atomic clocks that maintains global time standards. The world's most precise atomic clocks, located at national standards laboratories around the world, must all account for their specific gravitational environments to maintain synchronization.

Clocks at the U.S. Naval Observatory in Washington, D.C., the National Institute of Standards and Technology in Boulder, Colorado, and the International Bureau of Weights and Measures near Paris, France all tick at slightly different rates due to their different altitudes and local gravitational conditions. Maintaining coordinated universal time requires continuous corrections based on Einstein's gravitational time dilation formulas.

Even the precise GPS time system, which serves as a reference for countless technological applications, must account for gravitational time dilation effects at multiple levels. Not only do the satellite clocks need correction for their orbital altitude, but the ground-based monitoring stations that track and control the satellites are themselves at different altitudes and experience subtly different gravitational time dilation effects.

Laboratory Demonstrations

Modern physics laboratories routinely demonstrate gravitational time dilation using portable atomic clocks and modest elevation changes. In one famous demonstration, physicists at the National Institute of Standards and Technology measured time dilation by simply raising an optical atomic clock by half a meter on their laboratory bench.

The elevation change of just 50 centimeters was sufficient to create a measurable difference in the rate at which time passed. The raised clock gained about 1.6 parts in 10^{16} relative to the lower clock—an incredibly tiny effect that nevertheless represents a real difference in the flow of time itself.

Such experiments have been extended to detect time dilation effects in moving vehicles, elevators, and even from the thermal expansion of the laboratory building as it heats and cools throughout the day. These measurements demonstrate that gravitational time dilation is not just a cosmic phenomenon but an everyday reality that affects every moment of our lives, even if the effects are far too small for human senses to detect.

The Relativistic Geoid

The precise measurement of gravitational time dilation has practical applications beyond navigation and timekeeping. Since the effect depends on the local strength of gravity, atomic clocks can serve as extraordinarily sensitive gravitometers—instruments for mapping gravitational fields.

By comparing the rates of atomic clocks at different locations, scientists can map tiny variations in Earth's gravitational field with unprecedented precision. These measurements reveal the subtle effects of underground mass distributions, from oil and mineral deposits to groundwater aquifers and geological structures.

The concept of the "relativistic geoid"—a reference surface that accounts for gravitational time dilation effects—is becoming increasingly important for precise geodetic measurements. This surface represents locations where clocks would all tick at the same rate, providing a natural reference frame for understanding Earth's gravitational field and its effects on time.

Future applications may include using networks of atomic clocks to monitor changes in underground water levels, detect subtle crustal deformation that precedes earthquakes, or even track the motion of magma chambers beneath active volcanoes. The exquisite sensitivity of gravitational time dilation measurements opens new possibilities for understanding our planet's internal structure and dynamics.

Philosophical Implications

The reality of gravitational time dilation challenges our most fundamental intuitions about the nature of time. We naturally think of time as universal—the same everywhere, flowing at the same steady rate regardless of location or circumstance. But Einstein's theory and its experimental confirmation reveal a different reality: time is local, personal, and intimately connected to the gravitational environment.

This has profound implications for our understanding of simultaneity and causality. Events that appear simultaneous to an observer at sea level may not be simultaneous to an observer on a mountaintop, due to the different rates at which time flows at these different gravitational potentials. The very concept of "now" becomes relative and observer-dependent.

Yet remarkably, this strange behavior of time follows precise mathematical laws that allow us to calculate and predict exactly how clocks will behave in any gravitational environment. The universe may be stranger than our intuition suggests, but it operates according to principles that human reason can discover and understand.

Looking Toward the Horizon

As we've explored throughout this chapter, gravitational time dilation represents one of the most profound and well-tested consequences of Einstein's general theory of relativity. From the subtle effects measurable in Earth-based laboratories to the extreme time dilation near neutron stars and black holes, the phenomenon reveals that time itself is not the absolute, universal entity that Newton envisioned but rather a local phenomenon intimately connected to the curvature of spacetime.

The effects we've examined—from GPS satellites to atomic clocks on laboratory benches—demonstrate that we live our lives constantly moving through regions where time flows at subtly different rates. Every time Dr. Chen took the elevator to the top of Tokyo Skytree, she was literally time traveling into a region where time flows faster than at ground level, aging slightly more rapidly during her visit to the observation deck.

But the gravitational time dilation effects we've explored in this chapter, remarkable as they are, represent just the beginning of spacetime's capacity for temporal distortion. In the extreme gravitational environments around neutron stars and black holes, Einstein's equations predict time dilation effects so dramatic that they stretch our understanding of causality, simultaneity, and the very nature of temporal reality.

As we prepare to explore these extreme environments in upcoming chapters, we carry with us the profound realization that time is not just affected by gravity—it is part of the very fabric of spacetime that gravity curves and distorts. The ticking of atomic clocks, the synchronization of GPS satellites, and even our own aging as we move through Earth's gravitational field all reflect the deeper truth that we live in a universe where space, time, and gravity are unified in ways that continue to reveal the mathematical beauty and physical strangeness of Einstein's cosmic vision.

The gravity well that surrounds every massive object is not just a metaphor for describing gravitational attraction—it represents a real distortion in the fabric of spacetime that reaches into the very mechanism of time itself, slowing its flow and revealing that the universe operates according to principles far more subtle and remarkable than our everyday experience would suggest.

PART IV: EXTREME ENVIRONMENTS

Chapter 10: Black Holes – Where Time Stands Still

Dr. Andrea Valdez floated in the observation pod of the research vessel *Event Horizon*, humanity's most advanced deep-space exploration craft, positioned at what mission controllers deemed a "safe" distance of 100 kilometers from Cygnus X-1—one of the first black holes ever discovered and confirmed. Through the reinforced viewing port, she could see nothing where the black hole should be—just an absolute void in space, darker than the cosmic background, outlined only by the brilliant accretion disk of superheated matter spiraling inexorably toward the point of no return.

But Andrea wasn't looking for the black hole itself—that was impossible, by definition. She was observing something far more extraordinary: the behavior of time itself in the most extreme gravitational environment in the known universe. Her instruments were locked onto a robotic probe that had been released three hours earlier, programmed to fall toward the black hole while transmitting a steady stream of timing signals back to the research vessel.

What Andrea was witnessing defied every intuition about the nature of time. As the probe fell deeper into the black hole's gravitational well, its transmitted signals were arriving at increasingly longer intervals. The probe's internal clock, which had been ticking in perfect synchronization with the ship's chronometer when released, was now running dramatically slower. Time itself was being stretched, compressed, and ultimately frozen by the black hole's immense gravitational field.

From Andrea's perspective, watching from her relatively safe distance, the probe appeared to slow down as it approached the event horizon—the boundary beyond which nothing, not even light, can escape. But more remarkably, according to her measurements, the probe would never quite cross that boundary. Time dilation would become so extreme near the event horizon that the probe would appear to freeze just at the edge, its final moments stretched across eternity.

This is the ultimate expression of gravitational time dilation—the phenomenon where the curvature of spacetime becomes so severe that time itself approaches a complete standstill, at least from the perspective of distant observers. Black holes represent the most extreme gravitational environments possible in our universe, places where Einstein's equations reveal time behaving in ways that challenge our deepest assumptions about the nature of temporal reality.

The Schwarzschild Solution

To understand how black holes create such extreme time dilation effects, we must first examine the mathematical description of spacetime around these objects. The simplest case—and the one that reveals the most fundamental aspects of black hole physics—is the Schwarzschild solution, derived by Karl Schwarzschild just months after Einstein published his field equations in 1915.

Schwarzschild was solving Einstein's equations for the spacetime geometry around a spherically symmetric mass—a situation that describes many astronomical objects to good approximation, including black holes that don't rotate or carry electric charge. His solution revealed something remarkable:

there exists a critical radius around any massive object where the curvature of spacetime becomes so extreme that it creates a boundary from which nothing can escape.

This critical radius, now known as the Schwarzschild radius or event horizon, is given by:

$$\mathrm{rs} = \frac{2\mathrm{GM}}{c^2}$$

Where G is the gravitational constant, M is the mass of the object, and c is the speed of light. For a black hole with the mass of our Sun, the Schwarzschild radius is approximately 3 kilometers. For Earth, it would be about 9 millimeters—a sphere small enough to hold in your hand, yet containing our entire planet's mass.

The Schwarzschild solution describes how spacetime curves around a black hole, and embedded within this geometry is the prescription for how time flows at different distances from the event horizon. The time dilation factor for an observer at rest at distance r from the black hole, as measured by a distant observer, is given by:

$$\sqrt{1 - \frac{\mathrm{rs}}{\mathrm{r}}}$$

This deceptively simple formula contains profound physics. As r approaches the Schwarzschild radius r_s, the quantity under the square root approaches zero, making the time dilation factor approach zero as well. This means that time, as measured by distant observers, slows down dramatically near the event horizon and appears to stop completely at the horizon itself.

The Journey Toward the Event Horizon

To understand the extreme time dilation effects around black holes, let's follow Andrea's probe on its journey toward Cygnus X-1. The black hole has a mass roughly 21 times that of our Sun, giving it a Schwarzschild radius of about 62 kilometers. The probe begins its journey from a distance of 1,000 kilometers—far enough that spacetime curvature is noticeable but not yet extreme.

At this initial distance, the time dilation factor is:

$$\sqrt{1-\frac{62}{1000}} = \sqrt{(0.938)} = 0.968$$

The probe's clocks are running about 3.2% slower than identical clocks on Andrea's research vessel. This is significant but not yet dramatic—over the course of an hour, the probe's clocks would lag behind by about 2 minutes.

As the probe falls closer, the effects become more pronounced. At 200 kilometers from the black hole:

$$\sqrt{1-\frac{62}{200}} = \sqrt{(0.69)} = 0.831$$

Now the probe's clocks are running about 17% slower. An hour on the probe corresponds to about 72 minutes for Andrea.

At 100 kilometers—still twice the Schwarzschild radius:

$$\sqrt{1-\frac{62}{100}} = \sqrt{(0.38)} = 0.616$$

The time dilation is now extreme. The probe's clocks are running at only 62% of the rate measured by distant observers.

What feels like an hour to instruments on the probe appears as 97 minutes to Andrea.

As the probe approaches the event horizon at 62 kilometers, the time dilation becomes infinite from the perspective of outside observers. The probe appears to slow down dramatically, its signals becoming increasingly redshifted and faint, until it seems to freeze just at the boundary of the event horizon.

The Paradox of Perspectives

Here we encounter one of the most profound puzzles in black hole physics: the dramatic difference between what outside observers see and what someone falling into a black hole would experience. From Andrea's perspective on the research vessel, the probe never quite crosses the event horizon—it appears to slow down and freeze just at the boundary, its final moments stretched across eternity.

But what would the experience be like for a hypothetical astronaut riding along with the probe? Remarkably, from the falling observer's perspective, there's nothing special about crossing the event horizon. Their clocks continue to tick normally, they experience no unusual passage of time, and they cross the boundary in a finite amount of their own time.

This isn't a contradiction—it's a profound illustration of the relativity of time in extreme gravitational fields. The astronaut's proper time—the time measured by clocks in their own reference frame—flows normally throughout the journey. The extreme time dilation is only apparent when comparing the astronaut's time to that of distant observers.

This dual perspective reveals something fundamental about the nature of black holes and spacetime. The event horizon isn't a physical surface with special properties—it's a coordinate-dependent boundary that depends on the perspective of the observer. From the inside, spacetime continues smoothly across the horizon. From the outside, the horizon represents a boundary where time appears to stop.

Frame Dragging and Rotating Black Holes

Real black holes in nature are more complex than the simple Schwarzschild solution suggests. Most black holes rotate, often at tremendous speeds, and this rotation creates additional effects on spacetime geometry and time dilation. These rotating black holes are described by the Kerr solution, named after Roy Kerr who derived it in 1963.

Rotating black holes exhibit a phenomenon called frame dragging—the tendency for the spinning mass to "drag" spacetime along with it as it rotates. This creates a swirling pattern in spacetime around the black hole, similar to how a spoon stirring honey creates swirling currents in the liquid.

Frame dragging has profound implications for time dilation around rotating black holes. The time dilation factor becomes more complex, depending not only on the distance from the black hole but also on the angle relative to the rotation axis. Near the equatorial plane of a rapidly spinning black hole, frame dragging can actually reduce the time dilation compared to a non-rotating black hole of the same mass.

The mathematics becomes more complex for rotating black holes, but the physical insight remains the same: the curvature of spacetime, now enhanced by rotation, continues to stretch and compress time in ways that depend on both position and orientation relative to the spinning mass.

The Ergosphere and Time

Rotating black holes possess a unique region called the ergosphere—a zone outside the event horizon where spacetime is dragged around so strongly that nothing can remain at rest relative to distant observers. In this region, time dilation effects become even more bizarre and counterintuitive.

Within the ergosphere, any object must rotate with the black hole to some degree. For an object to remain "stationary" (not rotating with respect to distant stars), it would need to move faster than the speed of light, which is impossible. This creates a region where the very concept of being at rest becomes meaningless—everything must participate in the rotation of spacetime itself.

The ergosphere extends from the event horizon out to a distance that depends on the black hole's spin rate. For a maximally rotating black hole, the ergosphere can extend out to nearly twice the Schwarzschild radius. Within this region, time dilation effects are modified by the frame-dragging phenomenon, creating complex patterns of temporal distortion that depend on both radial distance and angular position.

Tidal Forces and Spaghettification

While time dilation represents the most dramatic temporal effect near black holes, the extreme curvature of spacetime also creates powerful tidal forces that have profound implications for any object approaching the event horizon. These tidal forces arise because different parts of an extended object experience different gravitational field strengths.

For an astronaut falling feet-first toward a black hole, the gravitational force on their feet (closer to the black hole) would be significantly stronger than the force on their head. This difference in gravitational force creates a stretching effect along the radial direction, while simultaneously compressing the astronaut in the perpendicular directions. This process, colorfully known as "spaghettification," would ultimately stretch the astronaut into a long, thin shape.

The strength of tidal forces depends on the mass of the black hole. Paradoxically, supermassive black holes—despite their enormous mass—have gentler tidal forces at their event horizons than stellar-mass black holes. This occurs because the Schwarzschild radius scales linearly with mass, while tidal forces scale as the inverse cube of distance. For supermassive black holes millions of times the mass of the Sun, the event horizon is so far from the central singularity that tidal forces remain manageable even at the boundary.

Hawking Radiation and Time

In 1974, Stephen Hawking made a revolutionary discovery about black holes that connected quantum mechanics with general relativity in an unexpected way. He showed that black holes are not completely black but instead emit thermal radiation—now known as Hawking radiation—due to quantum effects near the event horizon.

The temperature of this radiation is inversely proportional to the black hole's mass. A stellar-mass black hole with the mass of our Sun has a Hawking temperature of only 62 nanokelvins—far colder than the cosmic microwave background radiation. This means that stellar-mass black holes actually absorb more energy from the cosmic background than they emit, causing them to grow rather than shrink.

But for smaller black holes, Hawking radiation becomes significant. A black hole with the mass of a mountain would have a temperature of billions of degrees and would evaporate explosively in a fraction of a second. The time scale for black hole evaporation scales as the cube of the mass—larger black holes take exponentially longer to evaporate than smaller ones.

The connection to time dilation is subtle but profound. Hawking radiation emerges from quantum fluctuations in the curved spacetime near the event horizon, where time dilation effects are extreme. The radiation that distant observers detect as coming from the black hole actually originates from processes occurring in regions where time flows at vastly different rates compared to the outside universe.

Gravitational Waves from Black Holes

The 2015 detection of gravitational waves by LIGO opened a new window into black hole physics and provided unprecedented insights into how spacetime behaves in the most extreme gravitational environments. The first detected gravitational wave signal came from the merger of two black holes, each about 30 times the mass of the Sun, located over a billion light-years away.

During the final moments before merger, these black holes were orbiting each other at speeds approaching half the speed of light, experiencing time dilation effects that would have been extreme even by black hole standards. The gravitational waves themselves—ripples in spacetime—carry information about how time was flowing in these extreme conditions during the collision.

The LIGO detections confirm Einstein's predictions about how spacetime behaves when curved to the extreme. The gravitational wave signals match theoretical predictions based on Einstein's field equations with extraordinary precision, providing direct evidence that our understanding of time dilation and spacetime curvature remains valid even in the most extreme environments imaginable.

The Information Paradox

One of the deepest puzzles in black hole physics involves the relationship between time dilation, quantum mechanics, and information. According to quantum mechanics, information cannot be destroyed—if you know the complete quantum state of a system at one time, you can, in principle, calculate its state at any other time. But black holes seem to violate this principle.

When matter falls into a black hole, the information it carries appears to be lost forever, hidden beyond the event horizon where time, from the perspective of outside observers, has effectively stopped. Yet Hawking radiation appears to be purely thermal, carrying no information about what fell into the black hole. This creates the black hole information paradox—a fundamental conflict between general relativity and quantum mechanics.

The resolution of this paradox may require a deeper understanding of how time works near black holes. Some theories suggest that the extreme time dilation effects might play a crucial role in preserving information, perhaps allowing it to be slowly released through subtle correlations in Hawking radiation over the enormous time scales required for black hole evaporation.

Observational Evidence

While we cannot directly observe the extreme time dilation effects at black hole event horizons, astronomers have found compelling evidence for these effects in observations of matter spiraling into black holes. The hot gas in accretion disks around black holes emits X-rays with frequencies that depend on the motion and gravitational environment of the emitting material.

As this material spirals closer to the event horizon, the X-rays it emits become increasingly redshifted due to gravitational time dilation. The innermost stable circular orbit around a black hole—the closest distance at which matter can maintain a stable orbit—provides a natural laboratory for observing extreme time dilation effects.

Observations of these "relativistic iron lines" in X-ray spectra from black hole systems show exactly the gravitational redshift and time dilation effects predicted by Einstein's theory. The material in the innermost regions of the accretion disk experiences time dilation so extreme that its emission appears dramatically redshifted and time-stretched to distant observers.

The Holographic Principle

Recent theoretical developments in black hole physics have led to surprising insights about the nature of spacetime and information near event horizons. The holographic principle suggests that all the information about the interior of a black hole might be encoded on its two-dimensional surface—the event horizon itself.

This principle implies a deep connection between the extreme time dilation near black holes and the fundamental nature of space, time, and information. If the holographic principle is correct, then the apparent freezing of time at the event horizon might not represent a loss of information but rather a different way of encoding and preserving it.

These ideas push the boundaries of our understanding of spacetime itself. They suggest that the extreme time dilation effects we observe near black holes might be manifestations of even deeper principles governing the nature of reality—principles that connect gravity, quantum mechanics, and information in ways we're only beginning to understand.

Wormholes and Time Travel

The extreme curvature of spacetime near black holes has led physicists to speculate about even more exotic possibilities, including wormholes—hypothetical tunnels through spacetime that could connect distant regions of the universe. While purely theoretical, these structures would involve spacetime curvatures even more extreme than those found near black holes.

If wormholes exist, the time dilation effects associated with them could, in principle, allow for time travel. An astronaut who traveled through a wormhole might emerge at a different location and time than expected, due to the complex interplay of spacetime curvature and time dilation effects along the journey.

While such scenarios remain highly speculative, they illustrate how the extreme time dilation effects we observe near black holes point toward even stranger possibilities in the depths of Einstein's universe. The mathematics of general relativity, when pushed to its limits, reveals a cosmos where time itself becomes malleable in ways that challenge our most fundamental concepts about causality and temporal order.

The Ultimate Cosmic Laboratories

Black holes represent the ultimate natural laboratories for studying extreme physics. Their immense gravitational fields create conditions that cannot be replicated on Earth or in any conceivable human-built apparatus. The time dilation effects near black holes are so extreme that they provide tests of general relativity in regimes where the theory's predictions are pushed to their absolute limits.

Every observation of black holes—from the X-ray emission of their accretion disks to the gravitational waves produced by their mergers—confirms Einstein's century-old insights about the nature of space, time, and gravity. These observations reveal that even in the most extreme environments imaginable, where spacetime is curved almost beyond comprehension, Einstein's equations continue to describe reality with extraordinary precision.

The study of black holes has also revealed deep connections between gravity, thermodynamics, and quantum mechanics that continue to drive theoretical physics forward. The extreme time dilation effects near event horizons serve as a bridge between these different areas of physics, suggesting that our understanding of space and time is still evolving.

Looking into the Abyss

As Dr. Valdez continued to monitor her probe's approach to Cygnus X-1, she contemplated the profound implications of what she was observing. The probe's signals, now arriving at ever-increasing intervals due to extreme gravitational time dilation, provided a direct window into one of the universe's most fundamental mysteries: the nature of time itself in the presence of infinite curvature.

From her perspective, the probe would never quite cross the event horizon—it would appear to freeze at the boundary, its final moments stretched across eternity by the extreme curvature of spacetime. Yet she knew that from the probe's own perspective, if it could somehow report back, the crossing would occur in a finite time, with no unusual effects noticed by instruments falling with it.

This duality—the freezing of time for distant observers versus the normal flow of time for falling objects—captures something profound about the nature of black holes and spacetime itself. These objects represent regions where the familiar concepts of space and time break down, revealing the universe to be stranger and more beautiful than everyday experience would suggest.

The extreme time dilation effects around black holes stand as monuments to the power of human reason to comprehend the universe. Einstein's mathematical insights, derived from pure thought about the nature of space and time, have proven capable of describing the behavior of the most extreme objects in the cosmos. The equations he developed to understand the simple equivalence of gravity and acceleration have revealed temporal effects so dramatic that they challenge our deepest intuitions about the nature of reality itself.

As we prepare to explore other extreme environments where gravitational time dilation reaches remarkable levels, we carry with us the profound realization that black holes represent the ultimate expression of Einstein's insights about space, time, and gravity. In these cosmic laboratories, where spacetime curvature approaches infinity, time itself reveals its most exotic and mysterious properties, confirming that we live in a universe far stranger and more wonderful than our everyday experience would ever suggest.

The journey to understand black holes and their effects on time has taken us to the very edge of human knowledge, where the mathematics of general relativity meets the deepest questions about the nature of existence itself. In peering into these cosmic abysses, we've discovered that time—the most fundamental aspect of our experience—is far more malleable, mysterious, and magnificent than anyone had dared to imagine.

Chapter 11: Neutron Stars and White Dwarfs

Commander Jessica Park stood in the control room of the Atacama Large Millimeter Array, watching as the world's most sensitive radio telescopes turned their attention toward a seemingly empty patch of sky in the southern constellation Musca. To the naked eye, there was nothing remarkable about this region—just the familiar darkness of space punctuated by distant stars. But hidden within that darkness lay one of the most extraordinary laboratories for studying time dilation ever discovered by human science.

Twenty years of patient observation had revealed that this empty-seeming region contains PSR J1141-6545, a binary system consisting of two of the most exotic objects in the universe: a neutron star and a white dwarf locked in a cosmic dance that takes them around each other every 4.7 hours. What makes this system remarkable isn't just the extreme nature of these objects, but the fact that their gravitational effects are so intense that they provide a natural laboratory for observing time dilation effects that dwarf anything we can create on Earth.

As Jessica monitored the incoming data, she was witnessing something that would have seemed impossible to scientists just decades ago: the ability to measure how rapidly spinning massive objects literally drag spacetime around with them, creating swirling eddies in the fabric of reality that affect the flow of time itself. The neutron star in this system serves as an incredibly precise cosmic clock, sending out radio pulses with the regularity of the finest atomic timepieces, while its white dwarf companion creates gravitational time dilation effects so dramatic that they alter the very arrival times of those pulses in measurable ways.

This is the universe of neutron stars and white dwarfs—objects so dense and gravitationally powerful that they compress matter to unimaginable densities and stretch time in ways that reveal the true extremes of Einstein's relativistic universe.

White Dwarfs: Stellar Remnants in Miniature

White dwarf stars represent one of the most common endpoints of stellar evolution, yet they create gravitational environments unlike anything we encounter in our everyday experience. When a star like our Sun exhausts its nuclear fuel, it undergoes a dramatic transformation that compresses roughly the entire mass of the Sun into a sphere about the size of Earth.

The resulting white dwarf has a density so extreme that a teaspoon of its material would weigh as much as an elephant on Earth. This incredible compression creates gravitational fields at the white dwarf's surface that are typically 10,000 to 100,000 times stronger than Earth's gravity. Such intense gravitational fields produce correspondingly dramatic time dilation effects that can be observed and measured from Earth.

Consider a white dwarf with the mass of our Sun compressed into Earth's radius. At its surface, the gravitational time dilation factor becomes approximately 0.9995—meaning that clocks on the white dwarf's surface run about 0.05% slower than clocks in free space far from any gravitational influence. While this might seem like a small effect, it represents a dramatic departure from the weak gravitational time dilation we experience on Earth, where the effect is measured in parts per billion rather than parts per thousand.

This 0.05% time dilation means that for every year that passes for a distant observer, someone on the white dwarf's surface would experience about 2.6 days less time. Over the course of a century, the difference would accumulate to more than seven months—a substantial fraction of a human lifetime lost to the slower passage of time in the intense gravitational field.

The white dwarf in the PSR J1141-6545 system that Jessica was observing presents an even more extreme case. This particular white dwarf has been spun up to rotate once every 100 seconds by accreting matter from its companion star before that companion exploded to form the neutron star. This rapid rotation creates additional effects beyond simple gravitational time dilation—it literally drags spacetime around with it, creating the phenomenon known as frame dragging.

The Physics of Degenerate Matter

To understand how white dwarfs and neutron stars create such extreme time dilation effects, we must first appreciate the exotic physics of the matter they contain. Both objects are supported against gravitational collapse by quantum mechanical effects that have no analogue in our everyday experience.

White dwarfs are supported by electron degeneracy pressure—a quantum mechanical effect that prevents electrons from being compressed beyond a certain limit. As gravity tries to crush the white dwarf further, the electrons resist compression not because of any classical force, but because the Pauli exclusion principle forbids multiple electrons from occupying the same quantum state. This creates a pressure that can balance even the enormous gravitational forces trying to compress the star further.

But there's a limit to how much mass electron degeneracy pressure can support. The Chandrasekhar limit, calculated by the young Indian astrophysicist Subrahmanyan Chandrasekhar in the 1930s, shows that white dwarfs cannot exceed about 1.4 times the mass of the Sun. Above this critical mass, gravity overwhelms electron degeneracy pressure, and the white dwarf undergoes catastrophic collapse.

When this happens, the electrons are forced to combine with protons to form neutrons, creating an object supported by neutron degeneracy pressure—a neutron star. The gravitational fields around these objects are so intense that they create time dilation effects that make white dwarfs seem tame by comparison.

Neutron Stars: The Ultimate Dense Matter

Neutron stars represent the most extreme form of matter that can exist without collapsing into a black hole. These objects pack 1.4 to 2.2 times the mass of our Sun into a sphere only about 20 kilometers in diameter—roughly the size of a large city. The resulting density is so extreme that a teaspoon of neutron star material would weigh about 6 billion tons, roughly equal to the mass of Mount Everest.

The gravitational field at the surface of a typical neutron star is about 200 billion times stronger than Earth's gravity. Such intense fields create time dilation effects that are dramatic by any standard. At the surface of a neutron star, time runs approximately 15-30% slower than it does in free space, depending on the neutron star's exact mass and radius.

This means that for every hour experienced by someone on a neutron star's surface, between 1.15 and 1.4 hours would pass for a distant observer. The effect is so strong that it becomes easily observable in the electromagnetic radiation emitted by these objects. Light climbing out of a neutron star's gravitational well loses energy, appearing redshifted to distant observers—a direct manifestation of the gravitational time dilation occurring at the neutron star's surface.

The interior structure of neutron stars reveals even more exotic physics. At the core, matter exists in states that cannot be replicated in any earthly laboratory. The normal structure of atomic nuclei breaks down under the extreme pressure, creating a fluid of neutrons with a sprinkling of protons and electrons. At even higher densities, physicists theorize that more exotic particles might appear—strange quarks, kaons, pions, or even more speculative forms of matter that push the boundaries of our understanding of physics.

Pulsars: Cosmic Lighthouses and Precision Clocks

Many neutron stars manifest themselves as pulsars—rapidly rotating neutron stars that emit beams of radiation from their magnetic poles. As the neutron star rotates, these beams sweep across space like lighthouse beams, creating regular pulses of radiation that reach Earth with clockwork precision.

The pulsar in the PSR J1141-6545 system that Jessica was monitoring rotates 2.5387230404 times per second—a rate known to extraordinary precision through decades of careful timing measurements. This incredible regularity makes pulsars some of the most precise clocks in the universe, allowing astronomers to measure tiny changes in their orbital motion and timing that reveal relativistic effects.

But the precision of pulsar timing also means that these objects are exquisitely sensitive to gravitational time dilation effects. Any change in the gravitational environment around a pulsar—whether due to its own motion through curved spacetime or the gravitational influence of a companion object—will alter the rate at which time passes for the pulsar and therefore affect the arrival times of its pulses at Earth.

In binary systems like PSR J1141-6545, this sensitivity allows astronomers to map the gravitational fields of both objects with remarkable precision. The white dwarf's gravitational field affects the pulsar's motion and the flow of time in its vicinity, creating subtle but measurable changes in pulse arrival times that reveal the strength and structure of the gravitational time dilation throughout the system.

Frame Dragging: When Rotation Twists Spacetime

One of the most remarkable discoveries about the PSR J1141-6545 system came from analyzing two decades of pulsar timing data. Astronomers found that the entire orbit of the two stars was slowly tumbling in space—a phenomenon that could only be explained by frame dragging caused by the white dwarf's rapid rotation.

Frame dragging, predicted by Einstein's general relativity, occurs when a rotating massive object literally drags spacetime around with it as it spins. The effect is incredibly subtle around slowly rotating objects like Earth, but becomes dramatic around rapidly spinning compact objects like the white dwarf in this system.

The white dwarf completes one rotation every 100 seconds, spinning much faster than typical white dwarfs, which usually rotate once every few hours. This rapid rotation creates a swirling pattern in spacetime that affects the motion of nearby objects—in this case, the neutron star orbiting around it.

Over the 20 years since the system was discovered, this frame dragging has caused the pulsar's orbit to precess by about 150 kilometers. This orbital precession is detectable because it changes the Doppler shift of the pulsar's signals as the neutron star moves toward and away from Earth at slightly different velocities than expected from a simple elliptical orbit.

The measurement of frame dragging in this system represents the first detection of this relativistic effect in a binary star system, providing direct observational confirmation of one of Einstein's most subtle predictions about how rotating matter affects the geometry of spacetime.

The Formation History Written in Time

The extreme time dilation effects and frame dragging observed in the PSR J1141-6545 system tell a remarkable story about the formation and evolution of this binary pair. Unlike most pulsar systems, where the neutron star formed first, in this system the white dwarf is older than the pulsar—a configuration that requires a complex evolutionary history.

The most likely scenario begins with two normal stars orbiting each other. The more massive star evolved faster, eventually becoming a white dwarf while its companion remained a normal star. During this phase, matter from the still-evolving companion flowed onto the white dwarf, causing it to spin up to its current rapid rotation rate.

Later, the companion star exploded as a supernova, forming the neutron star we observe today as a pulsar. Remarkably, this violent explosion didn't disrupt the binary system, despite the fact that supernova explosions typically destroy binary orbits. The survival of this system through such a catastrophic event makes it one of the rarest types of binary systems known to astronomy.

The rapid rotation of the white dwarf, revealed through frame-dragging measurements, provides direct evidence for this evolutionary scenario. The spin-up process that occurred when matter flowed from the companion star left its signature in the form of enhanced frame dragging that can still be detected millions of years later.

Millisecond Pulsars: Precision Timekeepers

Some neutron stars have been spun up to even more extreme rotation rates through the same accretion process that accelerated the white dwarf in PSR J1141-6545. These millisecond pulsars rotate hundreds of times per second—rates so fast that the neutron star's equator moves at a significant fraction of the speed of light.

The fastest known pulsar, PSR J1748-2446ad, rotates 716 times per second. At this rate, the neutron star's equator is moving at about 24% of light speed, creating conditions where both gravitational and velocity-based time dilation effects become extreme. The combination of intense gravity and rapid rotation makes these objects natural laboratories for testing Einstein's theories under the most extreme conditions possible short of black holes.

The precision timing of millisecond pulsars has enabled some of the most stringent tests of general relativity ever performed. The orbital decay of binary pulsars due to gravitational wave emission matches Einstein's predictions to better than 0.1% accuracy, providing indirect evidence for the existence of gravitational waves decades before their direct detection by LIGO.

Neutron Star Mergers: Cosmic Laboratories

When two neutron stars spiral into each other and merge, they create gravitational time dilation effects that reach the very limits of what matter can endure before forming a black hole. These merger events, detected through gravitational waves and electromagnetic observations, provide glimpses into physics at the boundary between neutron stars and black holes.

During the final moments before merger, the two neutron stars experience tidal forces so extreme that they deform each other significantly. The gravitational time dilation in the region between the stars becomes so severe that time nearly stops from the perspective of distant observers, while matter is compressed and heated to temperatures exceeding those in the centers of the most massive stars.

The 2017 detection of gravitational waves from the neutron star merger GW170817 provided unprecedented insights into these extreme conditions. The gravitational wave signal revealed how matter behaves when compressed to densities exceeding those in atomic nuclei, while the accompanying gamma-ray burst and optical transient showed how the merger ejects heavy elements into space.

These observations confirmed that neutron star mergers are the primary source of many heavy elements in the universe, including gold and platinum. The extreme gravitational time dilation during these events affects the nuclear processes that create these elements, influencing their formation rates and the chemical evolution of the universe.

Measuring the Unmeasurable

The study of neutron stars and white dwarfs has pushed the boundaries of observational astronomy, requiring techniques that can measure incredibly tiny effects with extraordinary precision. Pulsar timing arrays, consisting of networks of radio telescopes monitoring dozens of pulsars simultaneously, can detect changes in pulse arrival times as small as a few nanoseconds.

These measurements are so precise that they must account for numerous relativistic effects, including gravitational time dilation, frame dragging, and the propagation of radio waves through curved spacetime. The analysis requires Einstein's field equations in their full complexity, as the simplified approximations that work near Earth are insufficient for these extreme gravitational environments.

The precision of pulsar timing has reached the point where it can detect gravitational waves directly by observing correlated changes in pulse arrival times across multiple pulsars. This technique, known as pulsar timing arrays, complements ground-based gravitational wave detectors by probing different frequency ranges and potentially detecting gravitational waves from supermassive black hole mergers.

The Future of Dense Matter Physics

As observational techniques continue to improve, neutron stars and white dwarfs will provide even more stringent tests of Einstein's theory and deeper insights into the nature of matter under extreme conditions. Future space-based X-ray telescopes will measure gravitational redshifts from neutron star surfaces with unprecedented precision, while improved pulsar timing will detect ever subtler relativistic effects.

The study of these objects also connects to fundamental questions about the nature of matter itself. The exact equation of state of matter at nuclear densities remains uncertain, with implications for understanding how many protons and neutrons can be packed together before gravity overwhelms quantum mechanical resistance and forms a black hole.

Recent observations of particularly massive neutron stars, some approaching twice the mass of the Sun, have pushed theoretical models to their limits. These observations suggest that neutron star matter may be even stiffer—more resistant to compression—than previously thought, with implications for understanding exotic forms of matter that may exist in neutron star cores.

Living Laboratories of Extreme Physics

What makes neutron stars and white dwarfs so valuable for understanding gravitational time dilation is their accessibility to observation despite their extreme nature. Unlike black holes, where the most interesting physics occurs beyond the event horizon where we cannot observe it directly, these objects allow us to study matter and spacetime under extreme conditions while still being able to detect radiation from their surfaces.

The white dwarf in PSR J1141-6545, spinning every 100 seconds and dragging spacetime around with it, provides a natural laboratory for studying frame dragging that would be impossible to create artificially. The neutron star, serving as an incredibly precise clock, allows astronomers to measure these effects with extraordinary accuracy over timescales of decades.

Similarly, isolated neutron stars provide laboratories for studying matter at densities that exceed those in atomic nuclei by factors of several. The gravitational time dilation at their surfaces creates observable effects in their thermal emission, allowing astronomers to measure the relationship between mass and radius that determines the neutron star equation of state.

The Cosmic Perspective

As Commander Jessica Park concluded her observations of PSR J1141-6545, she reflected on the remarkable fact that this system, hundreds of light-years from Earth, provides insights into the fundamental nature of space, time, and matter that cannot be obtained any other way. The gravitational time dilation effects around these objects are so extreme that they bridge the gap between the familiar physics of our solar system and the exotic conditions near black holes.

The white dwarf and neutron star in this system represent two different endpoints of stellar evolution, both creating gravitational environments where time itself behaves in ways that challenge our intuition while confirming the mathematical beauty of Einstein's equations. The frame dragging detected in this system provides direct evidence that massive objects really do twist spacetime around them as they rotate, exactly as Einstein predicted more than a century ago.

These observations remind us that we live in a universe filled with natural laboratories that push the boundaries of physics far beyond what we can achieve on Earth. Every pulsar pulse that reaches our radio telescopes carries information about how time flows in gravitational fields billions of times stronger than Earth's. Every measurement of gravitational redshift from a white dwarf surface provides a direct test of how gravity affects the passage of time.

The study of neutron stars and white dwarfs has revealed that extreme gravitational time dilation is not just a theoretical curiosity but an observable phenomenon that shapes the evolution of matter and energy throughout the universe. From the formation of heavy elements in neutron star mergers to the precision timing of cosmic clocks, these effects influence processes that determine the chemical composition of planets and the structure of space and time itself.

As we prepare to explore even more exotic environments where gravitational effects reach their ultimate extremes, we carry with us the profound realization that neutron stars and white dwarfs represent beacons of extreme physics that illuminate the deepest properties of Einstein's universe. These cosmic laboratories continue to reveal new aspects of how matter, energy, space, and time interact under conditions that stretch our understanding to its limits while confirming the remarkable power of human reason to comprehend the universe's most extreme phenomena.

In the end, the story of neutron stars and white dwarfs is a testament to the unity of physics across scales and environments. The same equations that describe the gentle time dilation experienced by GPS satellites also govern the extreme temporal effects around these compact objects, revealing a universe where Einstein's insights about the nature of spacetime apply from the most mundane technological applications to the most exotic astrophysical phenomena.

Chapter 12: The Early Universe

Dr. Elena Rodriguez adjusted the controls on the James Webb Space Telescope's spectral analysis software, her heart racing with the same excitement she had felt as a graduate student decades earlier when she first glimpsed the cosmic microwave background in her data. But tonight, she wasn't looking at the afterglow of the Big Bang—she was peering even deeper into cosmic history, analyzing light from quasars that had been racing toward Earth for over 12 billion years, carrying with it a message about the very nature of time itself in the infant universe.

The data streaming across her monitor told an extraordinary story. These ancient quasars—supermassive black holes feeding voraciously at the centers of the universe's first galaxies—were flickering and changing in their brightness patterns, but not at the rate she would expect. Instead, every fluctuation, every burst of activity, every change in luminosity was unfolding at exactly one-fifth the speed it would if these same objects existed in today's universe.

Elena was witnessing direct evidence of one of Einstein's most profound predictions: that time itself flowed differently in the early universe. Not because the fundamental nature of time had changed, but because the very fabric of spacetime was so dramatically different in that ancient epoch that it stretched every moment, every process, every physical phenomenon into a cosmic slow-motion dance that continued for billions of years.

This wasn't just an abstract theoretical prediction—it was a measurable, observable consequence of living in an expanding universe where space and time are woven together into the dynamic fabric of spacetime. Elena's observations, part of a groundbreaking study published in Nature Astronomy, represented the first direct detection of cosmological time dilation in the early universe, confirming that Einstein's insights about the nature of space and time apply not just to local phenomena but to the evolution of the cosmos itself.

The Expanding Spacetime Canvas

To understand how time could flow differently in the early universe, we must first appreciate that the universe itself—the very fabric of spacetime—has been expanding since the Big Bang. This isn't just an expansion of matter through space, like debris flying apart from an explosion. Instead, it's the expansion of space itself, carrying matter along with it like raisins in rising bread dough.

This expansion has profound implications for the nature of time. Einstein's general relativity reveals that space and time are not separate entities but aspects of a unified spacetime fabric. When space expands, it stretches not just the distances between objects but also the temporal intervals between events. This cosmic stretching of time creates the phenomenon known as cosmological time dilation.

The effect is similar to the gravitational time dilation we explored in previous chapters, but its origin is different. While gravitational time dilation arises from the curvature of spacetime near massive objects, cosmological time dilation emerges from the expansion of spacetime itself. As light travels to us from distant objects, both the light waves and the time intervals between events at the source get stretched by the expansion of the universe.

The mathematics governing this effect is elegantly simple. In an expanding universe, time dilation increases with distance and therefore with lookback time. When we observe objects at cosmological distances, we see them as they were billions of years ago, and the expansion of spacetime between then and now causes all processes at the source to appear slowed down by a factor directly related to how much the universe has expanded since the light was emitted.

For the quasars Elena was studying, which existed when the universe was only about 1.5 billion years old—roughly one-tenth its current age—this expansion factor corresponds to a time dilation of exactly 5 to 1. Every second of activity in those ancient quasars appears to us as five seconds of slow-motion cosmic theater.

Quasars as Cosmic Clocks

The breakthrough that enabled Elena's observations came from recognizing that quasars could serve as standardized clocks for measuring time dilation across cosmic distances. Quasars are among the most energetic and variable objects in the universe—supermassive black holes with masses millions to billions of times that of our Sun, surrounded by swirling accretion disks of superheated matter that glow brighter than entire galaxies.

What makes quasars ideal cosmic timepieces is their intrinsic variability. As matter spirals into the central black hole, the accretion disk undergoes complex instabilities that cause the quasar's brightness to fluctuate on timescales ranging from hours to years. These fluctuations follow patterns that are characteristic of the physical processes occurring in the accretion disk—patterns that should be the same regardless of when in cosmic history the quasar existed.

By studying nearly 200 quasars spanning a range of cosmic distances and ages, researchers could compare the timescales of variability for quasars at different epochs. If time dilation affects the early universe as Einstein's theory predicts, then the most distant quasars should show variability patterns that are systematically slowed down compared to nearby quasars.

The challenge lay in disentangling genuine time dilation effects from other factors that might influence quasar variability. Different quasar masses, feeding rates, and environmental conditions could all affect the characteristic timescales of their fluctuations. Only by carefully analyzing large samples of quasars and using sophisticated statistical methods could researchers isolate the pure time dilation signal from these complicating factors.

The results were unambiguous. When the research team led by Professor Geraint Lewis from the University of Sydney and Dr. Brendon Brewer from the University of Auckland analyzed their data, they found exactly the time dilation signature predicted by Einstein's theory. The most distant quasars, observed as they were over 12 billion years ago, showed variability patterns that were stretched out by exactly the factor expected from cosmological time dilation.

The Infant Universe in Slow Motion

The discovery that time flowed five times more slowly in the early universe opens a remarkable window into the physical conditions that existed during the cosmos's infancy. When we observe these ancient quasars, we're not just seeing objects that are very far away—we're literally watching the universe when it was a fundamentally different place, governed by the same physical laws but operating under conditions that stretched every process into cosmic slow motion.

Consider what this means for our understanding of early cosmic evolution. The formation of the first stars, the growth of the first black holes, the assembly of the first galaxies—all of these processes unfolded in an environment where time itself flowed at one-fifth the rate we experience today. From our modern perspective, the early universe appears to have been caught in a temporal molasses, with every physical process proceeding at a dramatically reduced pace.

This time dilation affected every aspect of early cosmic evolution. The nuclear fusion processes that powered the first stars proceeded more slowly from our perspective, though they felt perfectly normal to any hypothetical observer present at the time. The accretion of matter onto growing black holes, the collision and merger of early galaxies, the formation of the first heavy elements—all of these fundamental processes that shaped the architecture of the modern universe were stretched out by the expansion of spacetime itself.

The implications extend beyond individual astrophysical processes to the very structure of cosmic history. The epochs we identify in the early universe—the era of recombination when atoms first formed, the period of reionization when the first stars lit up the cosmos, the formation of the first galactic structures—all of these milestones in cosmic evolution were embedded in a temporal framework that flowed at a radically different rate than the time we experience today.

The Cosmic Microwave Background and Time

The cosmic microwave background (CMB)—the afterglow of the Big Bang itself—provides another perspective on time dilation in the early universe. This radiation, which we observe today as a faint glow filling all of space, was emitted when the universe was only about 380,000 years old and has been traveling toward us ever since, its wavelength stretched by the expansion of spacetime.

The CMB represents our most direct observational link to the early universe, but it also embodies the time dilation effects we've been discussing. The radiation we detect today has been redshifted by a factor of about 1,100, meaning the universe has expanded by this factor since the CMB was emitted. This same expansion factor implies that time dilation effects in the CMB era would have been even more extreme than those observed in the quasar studies.

If we could somehow observe processes occurring at the time of CMB emission, they would appear to unfold at roughly 1/1100th the rate of equivalent processes today. The formation of the first atoms, the decoupling of matter and radiation, the subtle density fluctuations that would eventually grow into galaxies and galaxy clusters—all of these epoch-defining events occurred in a temporal framework stretched to almost incomprehensible proportions by the expansion of spacetime.

The discovery of time dilation in early quasars provides observational confirmation that these extreme time dilation effects really did characterize the early universe. It validates our understanding of how the expansion of spacetime affects not just the propagation of light but the very flow of time itself across cosmic history.

Dark Energy and Accelerating Expansion

The time dilation observed in early quasars also provides insights into one of the most mysterious components of our universe: dark energy. This enigmatic form of energy, which appears to be causing the expansion of the universe to accelerate, might seem like it should complicate the time dilation picture. However, the research reveals something remarkable about the fundamental nature of cosmological time dilation.

As Professor Lewis emphasized in describing the discovery, "This slowing down of time doesn't care about things like dark energy, dark matter, or any of the stuff in the universe. It is a property of spacetime itself". The time dilation effect emerges purely from the expansion of spacetime and would occur regardless of what causes that expansion or how the expansion rate changes over cosmic time.

This universality of cosmological time dilation provides a powerful tool for understanding cosmic evolution. While dark energy affects the rate at which the universe expands, and therefore influences how much total expansion occurs between different epochs, the time dilation effect depends only on the total amount of expansion that has occurred, not on the specific mechanisms driving that expansion.

The confirmation of time dilation in early quasars thus provides support for our standard model of cosmology, including the existence of dark energy. The observed time dilation matches exactly what would be predicted based on our understanding of how the universe has expanded since the epoch when these quasars were active, confirming that our cosmological models correctly describe the evolution of spacetime itself.

The Formation of Cosmic Structure

Understanding time dilation in the early universe transforms our perspective on how cosmic structures formed and evolved. The galaxies, galaxy clusters, and large-scale structures we observe today all trace their origins back to tiny density fluctuations in the early universe—fluctuations that grew under the influence of gravity over billions of years.

But this growth didn't occur in a temporally uniform environment. The earliest phases of structure formation unfolded in an epoch when time itself flowed much more slowly than it does today. This has profound implications for understanding how quickly structures could form and how the various physical processes involved in galaxy formation played out over cosmic time.

Consider the formation of the first stars, which likely occurred when the universe was only a few hundred million years old. From our perspective today, these stars formed in an environment where time flowed perhaps 10 times more slowly than it does now. The nuclear fusion processes that powered these stars, the stellar winds that dispersed heavy elements into space, the supernova explosions that marked their violent deaths—all of these processes were embedded in a temporal framework that was dramatically stretched compared to equivalent processes occurring today.

This temporal stretching helps explain some puzzling observations about the early universe. Astronomers have discovered surprisingly massive galaxies and supermassive black holes at very high redshifts, corresponding to very early epochs in cosmic history. The time dilation effect means that these objects had more time to grow and evolve than simple calculations based on modern time flow would suggest, helping to resolve apparent contradictions between theory and observation.

Gravitational Waves and Time Dilation

The recent detection of gravitational waves by LIGO and other interferometric detectors adds another dimension to our understanding of time dilation in cosmic evolution. These ripples in spacetime itself carry information about violent events in the distant universe, including the merger of black holes and neutron stars.

When gravitational waves from distant sources reach Earth, they too are affected by cosmological time dilation. A black hole merger that occurred billions of years ago would appear to unfold more slowly than an equivalent event occurring in the nearby universe. The gravitational wave signal would be stretched in time, with all characteristic frequencies shifted to lower values by the expansion of spacetime.

This effect must be carefully accounted for when analyzing gravitational wave detections from distant sources. The observed duration and frequency evolution of a gravitational wave signal depends not only on the intrinsic properties of the merging objects but also on the cosmological time dilation factor corresponding to the source's distance and epoch.

Future gravitational wave detectors, with enhanced sensitivity to lower frequencies, may be able to detect black hole mergers from the very early universe, providing another avenue for studying time dilation effects during the cosmos's infancy.

These observations would complement the quasar studies by probing different types of astrophysical phenomena in the early universe, all united by their embedding in the stretched temporal framework of the expanding cosmos.

The Anthropic Perspective

The discovery of extreme time dilation in the early universe raises fascinating questions about the nature of time and observation. From our perspective in the modern universe, the early cosmos appears to have been trapped in slow motion, with every process unfolding at a dramatically reduced pace. But this perspective is necessarily subjective—it depends on our particular location in cosmic space and time.

If conscious observers had existed in the early universe (which they almost certainly did not, given the extreme conditions), they would have experienced time flowing normally from their own perspective. A hypothetical early-universe observer would have seen processes around them unfolding at normal rates, while any observations of the distant future (including our current epoch) would have appeared dramatically accelerated.

This relativity of temporal experience extends to all physical processes. The quasars Elena observed were not "actually" flickering slowly—they were flickering at their natural rates, as determined by the physical processes in their accretion disks. The slow-motion appearance is a consequence of observing these processes across the vast temporal and spatial gulf created by the expansion of spacetime.

This perspective highlights one of the most profound implications of Einstein's relativity: there is no universal or absolute time against which all events can be measured. Time is always local, always dependent on the observer's reference frame, and always relative to the gravitational and kinematic environment in which observations are made.

Technological Implications

The confirmation of cosmological time dilation has practical implications for astronomical observations and our understanding of cosmic evolution. When astronomers observe distant objects, they must now account for the fact that all processes appear slowed down by factors that depend on the object's redshift and epoch.

This has implications for understanding the intrinsic properties of distant galaxies, quasars, and other astronomical objects. Apparent timescales for variability, stellar formation rates, and evolutionary processes must all be corrected for cosmological time dilation to understand the true physics occurring at the source.

Future astronomical surveys, including those planned for next-generation telescopes, will incorporate these time dilation corrections from the outset. The Extremely Large Telescope, the Nancy Grace Roman Space Telescope, and other upcoming facilities will observe objects across the full range of cosmic time, requiring sophisticated models that account for how the expansion of spacetime affects the apparent temporal evolution of distant sources.

The discovery also validates the use of quasars and other variable sources as standardized clocks for cosmological studies. Just as Type Ia supernovae serve as "standard candles" for measuring cosmic distances, variable quasars can serve as "standard clocks" for measuring the time dilation effects that reveal the expansion history of the universe.

The Deep Connection to Cosmic Evolution

Understanding time dilation in the early universe reveals a deep connection between the fundamental nature of spacetime and the evolution of cosmic structures. The universe we observe today—with its rich hierarchy of galaxies, stars, and planets—emerged from simpler initial conditions through processes that were embedded in a dramatically different temporal framework.

The density fluctuations that eventually grew into galaxies experienced this temporal stretching during their early evolution. The first generation of stars, which synthesized the heavy elements essential for planet formation and life, lived and died in an epoch when time itself flowed at a fraction of its current rate. The supermassive black holes that now reside at the centers of galaxies grew from smaller seeds through accretion processes that unfolded in slow motion from our modern perspective.

This temporal dimension of cosmic evolution adds richness to our understanding of how the universe developed from its initial state to its current configuration. The processes that shaped cosmic structure didn't just occur in different locations and under different physical conditions—they occurred in a fundamentally different temporal environment, one where the expansion of spacetime stretched every moment and every process into a cosmic slow-motion ballet.

Looking Forward and Backward

As Elena completed her analysis of the quasar time dilation data, she reflected on the remarkable journey that had brought her to this moment. The light she was studying had been traveling toward Earth for over 12 billion years, carrying information about conditions in the universe when it was less than one-tenth its current age. Every photon in her dataset was a messenger from an epoch when time itself flowed at one-fifth its current rate.

But this discovery also points toward future possibilities. As astronomical technology continues to advance, we may be able to probe even deeper into cosmic history, observing time dilation effects from epochs even closer to the Big Bang. The James Webb Space Telescope and its successors may reveal how time flowed in the universe when it was only a few hundred million years old, when the first stars and galaxies were just beginning to form.

These future observations will test the limits of our understanding of spacetime and cosmic evolution. They may reveal new physics operating in the extreme conditions of the very early universe, or they may confirm that Einstein's insights about the nature of space and time apply universally across cosmic history.

The study of time dilation in the early universe also connects to fundamental questions about the nature of time itself. If time flowed differently in the past, what does this tell us about the relationship between time and physical processes? How does the subjective experience of temporal flow relate to the objective measurements of cosmological time dilation? These questions touch on some of the deepest problems in physics and philosophy.

The Cosmic Calendar Revisited

The discovery of cosmological time dilation requires us to revisit our understanding of the cosmic calendar—the timeline that chronicles the major events in the universe's history. When we say that the cosmic microwave background was emitted 380,000 years after the Big Bang, or that the first stars formed a few hundred million years later, we must remember that these durations are measured in our current temporal framework.

From the perspective of observers living through those early epochs, the subjective experience of time would have been normal. But from our modern vantage point, those early eras represent extended periods where every process unfolded in slow motion. The 380,000 years between the Big Bang and CMB emission would have felt like 380,000 years to any hypothetical observer present at the time, but it represents a much longer duration when viewed from our temporally accelerated perspective.

This temporal stretching affects our understanding of how much "time" was available for various processes in the early universe. The formation of the first stars, the growth of the first black holes, the assembly of the first galaxies—all of these processes had more subjective time to unfold than simple calculations based on modern time flow would suggest.

The realization transforms our perspective on cosmic evolution from a linear progression through time to a more complex journey through an evolving spacetime framework where the very nature of temporal flow changes as the universe expands and evolves.

The Unity of Einstein's Vision

Elena's discovery of time dilation in early quasars represents more than just another confirmation of Einstein's theory—it reveals the remarkable unity of his vision of spacetime. The same mathematical framework that describes time dilation for GPS satellites and falling elevators also governs the flow of time across cosmic history. The equations that predict gravitational time dilation near black holes also determine how time flows in the expanding universe.

This unity demonstrates the power of Einstein's geometric interpretation of gravity and spacetime. By recognizing that space and time form a unified fabric that can be curved by matter and energy, Einstein provided a framework that describes phenomena from the most local to the most cosmic scales. The time dilation effects observed in Elena's quasar data emerge from the same fundamental principles that govern the operation of atomic clocks and the navigation accuracy of GPS satellites.

The universality of these effects across such vastly different scales and environments testifies to the deep mathematical structure underlying physical reality. The universe operates according to geometric principles that human reason can discover and understand, revealing connections between phenomena that might otherwise seem completely unrelated.

As we continue to explore the implications of time dilation across different environments and epochs, we carry with us Einstein's profound insight that the universe is comprehensible, that its deepest principles can be expressed in mathematical form, and that these principles apply universally across space and time. The early universe, with its dramatically slowed temporal flow, represents just one more manifestation of the geometric beauty that underlies all of physical reality.

The story of time dilation in the early universe is ultimately a story about the nature of time itself—how it flows, how it responds to the expansion of space, and how it shapes the evolution of everything from individual stars to the cosmos as a whole. In peering back to the universe's infancy and observing time flowing in slow motion, we've gained new insights into one of the most fundamental aspects of existence while confirming once again that we live in Einstein's universe, where space and time dance together in ways more beautiful and stranger than our everyday experience could ever suggest.

PART V:

PRACTICAL IMPLICATIONS AND FUTURE FRONTIERS

Chapter 13: Navigation by Einstein

Captain Maria Santos checked her ship's position one final time as the cargo vessel *Meridian* approached the narrow entrance to the Port of Rotterdam. On her bridge display, a precise blue dot marked their location at 51.9244° North, 4.4777° East—accurate to within three meters despite traveling through some of the busiest shipping lanes in the world. Around her, dozens of other vessels navigated the complex approach using the same invisible constellation of guides: twenty-four satellites orbiting 20,200 kilometers above Earth, each one a testament to the practical genius of Albert Einstein.

What Captain Santos couldn't see—what none of the millions of people using GPS navigation that day could observe—was the constant, precise correction occurring twenty-four hours a day in those satellites' atomic clocks. Every second of every day, the onboard computers were adjusting their time measurements to account for the fact that Einstein's theories of relativity make time flow differently in space than it does on Earth's surface.

Without these corrections, the GPS constellation would become useless for navigation within hours. Captain Santos's precise position reading would drift by more than 10 kilometers every day. The blue dot that showed her ship's location would wander aimlessly across her chart, making precision navigation impossible. The entire $1.4 trillion global economy that depends on GPS—from shipping and aviation to rideshare apps and emergency services—would collapse.

This is navigation by Einstein: the most widespread, practical application of relativistic physics in human history, invisibly operating in the pockets and vehicles of billions of people who rely daily on corrections derived from thought experiments about the nature of space and time.

The Relativistic Challenge

To understand why GPS requires Einstein's equations to function, we must first appreciate the extraordinary precision demanded by satellite navigation. The basic principle is elegantly simple: measure the time it takes radio signals to travel from satellites to your receiver, calculate distances based on the speed of light, and use triangulation to determine your exact position.

But executing this principle requires timing precision that pushes the boundaries of measurement technology. GPS signals travel at the speed of light—299,792,458 meters per second—which means that light travels about 30 centimeters in one nanosecond (one billionth of a second). To achieve position accuracy of a few meters, GPS timing must be precise to within a few nanoseconds.

This extraordinary precision requirement means that even tiny timing errors have dramatic consequences. A timing error of just one microsecond—one millionth of a second—would cause position errors of about 300 meters. Errors of tens of microseconds would make GPS useless for any practical navigation purpose.

The problem is that GPS satellites don't exist in the same relativistic environment as receivers on Earth's surface. Each satellite experiences both gravitational and velocity-based effects that alter the rate at which time passes, creating systematic timing errors that would quickly destroy the system's accuracy without constant correction.

The Dual Nature of Relativistic Effects

GPS satellites experience two distinct relativistic effects that work in opposite directions. Understanding both is crucial to appreciating how navigation by Einstein actually works in practice.

Special Relativistic Time Dilation arises from the satellites' orbital velocity of approximately 14,000 kilometers per hour (3.87 kilometers per second). At these speeds, special relativity predicts that the satellites' clocks should run slower than identical clocks on Earth's surface. The effect is small but measurable: satellite clocks lose about 7 microseconds per day due to their high-speed motion through space.

General Relativistic Time Dilation comes from the weaker gravitational field at the satellites' orbital altitude. According to Einstein's equivalence principle and general relativity, clocks in weaker gravitational fields run faster than clocks in stronger fields. At 20,200 kilometers above Earth's surface, the gravitational field is significantly weaker than at sea level, causing satellite clocks to gain about 45 microseconds per day.

The net effect combines these two opposing influences: satellite clocks run about 38 microseconds per day faster than Earth-based clocks (45 microseconds gained from weaker gravity minus 7 microseconds lost from high velocity). This might seem like an insignificant amount, but it represents a timing error large enough to make GPS completely unreliable for navigation.

The Technical Solution

The GPS system's designers solved this relativistic timing problem through an elegant engineering approach that demonstrates the practical power of Einstein's insights. Rather than trying to correct for relativistic effects in software after the fact, they built the corrections directly into the satellites' hardware.

Each GPS satellite carries atomic clocks—devices so precise they would gain or lose less than one second over 300,000 years under normal circumstances. But instead of setting these clocks to run at their natural frequency of 10.23 megahertz, GPS engineers deliberately set them to run at 10.22999999543 megahertz—slightly slower than they would naturally run on Earth.

This pre-correction means that when the clocks are launched into their orbital environment, the relativistic effects cause them to speed up by exactly the right amount to keep perfect time with Earth-based reference clocks. The 38 microseconds per day of relativistic speedup brings the satellite clocks into synchronization with the ground-based time standard.

The precision of this correction is extraordinary. GPS engineers must account for relativistic effects to within nanoseconds, requiring calculations based on Einstein's field equations carried out to many decimal places. The system represents one of the most precise large-scale applications of general relativity ever implemented.

The Mathematics of Navigation

The mathematical foundation of GPS navigation reveals how Einstein's insights translate into practical positioning calculations. Each satellite broadcasts a timing signal that includes the precise time the signal was transmitted and the satellite's exact orbital position. A GPS receiver measures when each signal arrives and calculates the distance to each satellite using the simple formula: distance = speed of light × travel time.

But this calculation only works if the timing is absolutely precise. The "pseudorange" equation that GPS receivers use can be written as:

$$\rho = c(t_R - t_S) + c(\delta t_R - \delta t_S)$$

Where ρ is the measured distance, c is the speed of light, t_R and t_S are the reception and transmission times, and δt_R and δt_S are the clock errors in the receiver and satellite respectively.

The relativistic corrections ensure that δt_S , the satellite clock error—remains negligible. Without these corrections, the satellite clock errors would grow by 38 microseconds each day, causing position errors that increase by more than 10 kilometers daily.

To determine a position in three-dimensional space, a GPS receiver needs distance measurements from at least four satellites. The fourth measurement is necessary because GPS receivers don't carry atomic clocks—they use the fourth satellite measurement to solve for both position and the receiver's internal clock error simultaneously.

Recent Advances: The Galileo Achievement

Europe's Galileo satellite navigation system has recently provided the most precise test of Einstein's general relativity ever achieved using satellites. Two Galileo satellites launched into highly elliptical orbits—rather than the circular orbits used by standard navigation satellites—have enabled unprecedented measurements of gravitational time dilation.

These satellites experience significant variations in gravitational field strength as their orbits carry them from about 17,000 kilometers to 26,000 kilometers altitude. The changing gravitational environment creates measurable variations in the rate at which their atomic clocks tick, providing a natural laboratory for testing Einstein's predictions about gravitational time dilation.

Two independent physics teams analyzing more than 1,000 days of data from these satellites achieved about a fivefold improvement in measuring the accuracy of gravitational time dilation effects. Their results, published in Physical Review Letters, represent the first significant improvement in gravitational redshift measurements in more than 40 years.

The extraordinary precision of these measurements was enabled by the exceptional stability of Galileo's atomic clocks, the accuracy of orbit determination, and the presence of laser retroreflectors that allow ground-based measurements precise enough to separate clock effects from orbital errors.

The Global Infrastructure

The success of navigation by Einstein has created a global infrastructure that extends far beyond simple positioning. GPS time—maintained through constant relativistic corrections—serves as a critical reference for countless technological systems worldwide.

Financial markets use GPS time stamps to sequence high-frequency trading transactions. Power grids synchronize their operations using GPS timing signals. Telecommunications networks rely on GPS for the precise timing required for digital communications. Emergency services use GPS for dispatching and coordination. Even internet infrastructure depends on GPS timing for network synchronization.

This ubiquitous reliance on relativistically corrected timing has made GPS one of the most critical infrastructures in modern civilization. The system processes timing corrections based on Einstein's equations billions of times per day, ensuring that relativistic effects don't accumulate into system-destroying errors.

The economic impact is staggering. Studies estimate that GPS contributes more than $1.4 trillion annually to the U.S. economy alone, with similar benefits worldwide. Every smartphone navigation app, every rideshare pickup, every precision agriculture application, every aircraft approach—all depend on continuous relativistic corrections derived from Einstein's century-old insights about space and time.

Beyond Earth: Relativistic Navigation

As humanity expands beyond Earth, the lessons learned from GPS navigation by Einstein will become even more critical. Proposed missions to Mars, lunar bases, and deep space exploration will require navigation systems that account for even more complex relativistic effects.

Near massive objects like Jupiter or in the varying gravitational fields between planets, relativistic effects can become much more pronounced than those experienced in Earth orbit. Future interplanetary navigation systems will need to account for gravitational time dilation effects that change continuously as spacecraft travel between different gravitational environments.

The precision timing requirements for deep space navigation will also be more demanding. Communicating across interplanetary distances requires timing accuracy that accounts for both special and general relativistic effects as signals travel through the varying gravitational fields of the solar system.

The Sagnac Effect and Rotating Frames

GPS navigation must also account for a subtle relativistic effect known as the Sagnac effect, which arises from Earth's rotation. Because Earth rotates beneath the GPS satellites, the effective distances between satellites and ground receivers depend on the direction of signal propagation relative to Earth's rotation.

Signals traveling in the same direction as Earth's rotation effectively travel slightly farther than signals traveling in the opposite direction, because the receiver moves during the signal's transit time. This creates small but measurable timing differences that must be corrected to maintain navigation accuracy.

The Sagnac effect demonstrates how even Earth's rotation—something we rarely think about in daily life—creates relativistic corrections that are essential for precise navigation. GPS engineers must account for this effect using calculations derived from Einstein's special theory of relativity applied to rotating reference frames.

Military Origins and Civilian Applications

GPS was originally developed by the U.S. Department of Defense for military applications, but the relativistic precision required for military targeting also enabled unprecedented civilian navigation capabilities. The same atomic clocks and relativistic corrections needed for precision missile guidance also support civilian applications ranging from surveying to smartphone apps.

This dual-use nature of GPS demonstrates how fundamental physics research can have unexpected practical applications. Einstein's theoretical insights about the nature of space and time, developed through pure thought experiments, became essential for one of the most widespread technologies in human history.

The military precision requirements also drove the development of increasingly accurate atomic clocks and more sophisticated relativistic correction algorithms. Modern GPS satellites carry cesium and rubidium atomic clocks that maintain accuracy to within a few nanoseconds, enabling position accuracies that were unimaginable when the system was first conceived.

Precision Agriculture and Economic Impact

One of the most economically significant applications of navigation by Einstein occurs in precision agriculture, where GPS guidance systems enable farmers to plant, cultivate, and harvest crops with centimeter-level accuracy. This precision farming reduces waste, increases yields, and minimizes environmental impact by ensuring that fertilizers and pesticides are applied exactly where needed.

The relativistic corrections that make such precision possible have revolutionized modern agriculture. Tractors guided by GPS can plant seeds in perfectly straight rows, apply fertilizer with minimal overlap, and harvest crops with maximum efficiency. The economic benefits extend to reduced fuel consumption, decreased chemical use, and increased crop yields that help feed the world's growing population.

Scientific Applications

Beyond navigation and timing, GPS has become an essential tool for scientific research that depends on precise positioning and timing. Seismologists use GPS to measure tiny movements of Earth's crust, detecting earthquake precursors and monitoring tectonic plate motion with millimeter precision.

Atmospheric scientists use GPS signals to study the ionosphere and troposphere, extracting information about atmospheric water vapor, temperature, and electron density from the way GPS signals are affected as they pass through different atmospheric layers.

Geodesists use GPS to measure changes in Earth's shape and gravitational field, tracking everything from post-glacial rebound to the effects of large earthquakes on Earth's rotation. These applications all depend on the relativistic precision of GPS timing and positioning.

The Future of Relativistic Navigation

As navigation technology continues to advance, relativistic effects will become increasingly important. Next-generation atomic clocks promise even greater precision, enabling navigation systems that can detect and correct forever subtler relativistic effects.

Optical atomic clocks, which use laser light instead of microwave radiation, offer the potential for timing precision that exceeds current systems by orders of magnitude. These ultra-precise clocks could enable navigation systems sensitive enough to measure gravitational time dilation effects from local topography, groundwater variations, and other small-scale gravitational anomalies.

Future navigation systems may also incorporate quantum sensing techniques that exploit relativistic effects in new ways. Quantum gravimeters and quantum inertial sensors could complement satellite navigation by providing navigation capabilities that work even when satellite signals are unavailable.

Educational and Cultural Impact

The success of GPS has made Einstein's relativity theories more accessible to the general public than any previous application of advanced physics. Millions of people use devices that depend on relativistic corrections every day, creating opportunities to explain how fundamental physics research translates into practical technology.

Educational programs increasingly use GPS as an entry point for teaching relativity concepts, showing students how abstract theories about space and time have immediate practical relevance. The tangible nature of GPS navigation makes relativistic effects concrete and understandable in ways that pure theoretical discussions often cannot achieve.

Looking Back and Forward

As Captain Santos guided the *Meridian* safely into Rotterdam Harbor, she reflected on the invisible relativistic infrastructure that had made her precise navigation possible. The GPS constellation orbiting overhead was continuously executing billions of calculations based on Einstein's century-old insights, ensuring that timing errors didn't accumulate into navigation failures.

The success of navigation by Einstein represents more than just a technological achievement—it demonstrates the profound practical value of fundamental physics research. Einstein's thought experiments about falling elevators and accelerating trains, his insights about the equivalence of gravity and acceleration, and his mathematical description of curved spacetime have become essential components of modern civilization.

Every GPS-guided ship, aircraft, and vehicle represents a vindication of Einstein's faith that the universe operates according to mathematical principles that human reason can discover and understand. The atomic clocks orbiting Earth, continuously correcting for relativistic effects, stand as monuments to the power of theoretical physics to transform human capabilities.

But navigation by Einstein also points toward future possibilities. As we venture beyond Earth and develop ever more precise navigation systems, Einstein's insights about space and time will become even more essential. The relativistic effects that require careful correction in Earth orbit will become dominant factors in interplanetary navigation, requiring new applications of general relativity to even more extreme environments.

The story of GPS navigation demonstrates that Einstein's universe is not just a theoretical curiosity but the practical reality in which we live and work every day. The satellites orbiting overhead, the smartphones in our pockets, and the navigation systems that guide global commerce all depend on our understanding of how space, time, and gravity interact in the relativistic cosmos that Einstein revealed.

In learning to navigate by Einstein, humanity has not just solved a practical problem—we've confirmed once again that we live in a universe where space and time are malleable, where gravity shapes the flow of time itself, and where the deepest truths about reality can be discovered through human reason and applied to transform our technological capabilities. The next time you use GPS navigation, remember that you're relying on one of the most elegant and successful applications of Einstein's insights about the fundamental nature of space, time, and gravity—insights that continue to guide us through both the familiar territories of Earth and the vast cosmos beyond.

Chapter 14: Testing the Limits

Dr. Rajesh Patel stood in the pristine control room of the Laser Interferometer Gravitational-Wave Observatory (LIGO) in Livingston, Louisiana, watching data streams from the most sensitive measuring devices ever constructed by human civilization. Each of the facility's two perpendicular arms stretched 4 kilometers into the Louisiana countryside, containing laser interferometers so precise they could detect changes in distance smaller than 1/10,000th the width of a proton—a sensitivity that would notice if the distance to the nearest star changed by the width of a human hair.

On September 14, 2015, at 09:50:45 UTC, both LIGO detectors simultaneously registered a signal that would forever change physics and astronomy: the direct detection of gravitational waves from the collision of two black holes over a billion light-years away. But for Rajesh, watching the data analysis unfold in real-time, this detection represented something even more profound—the most extreme test of Einstein's general theory of relativity ever achieved, confirming the theory's predictions in gravitational fields so intense that they approached the very limits of what spacetime itself could endure.

The gravitational waves that LIGO detected had been created in an environment where time dilation effects were so extreme that they challenged the boundaries of physical law. In the final milliseconds before the black holes merged, each was orbiting the other at nearly half the speed of light, experiencing gravitational fields billions of times stronger than anything in our solar system, in regions where the curvature of spacetime approached the mathematical singularities that mark the breakdown of classical physics.

This is the frontier of testing Einstein's limits—pushing the boundaries of experimental precision to probe regimes where the effects of relativity become so extreme that they reveal new aspects of space, time, and gravity that no previous generation of scientists could have imagined studying.

The Classical Tests Revisited

Einstein's journey from theoretical insight to experimental validation began with three "classical tests" that he proposed in 1915 as definitive ways to distinguish his general theory of relativity from Newton's theory of gravity. These tests—the precession of Mercury's orbit, the bending of starlight by the Sun, and the gravitational redshift of light—established the experimental foundation for all subsequent tests of relativistic physics.

The 1919 solar eclipse expedition led by Arthur Eddington provided the first dramatic confirmation of Einstein's predictions. During the eclipse, when the Sun's bright disk was blocked by the Moon, Eddington's team photographed stars near the Sun's edge and measured how their apparent positions shifted due to the bending of their light by the Sun's gravitational field. The measured deflection of 1.75 arcseconds matched Einstein's prediction exactly, distinguishing it clearly from both the Newtonian prediction of 0.87 arcseconds and the possibility of no deflection at all.

This eclipse measurement represented extraordinary precision for its time, but it pales in comparison to modern tests. Today's radio interferometry can measure the positions of distant quasars with precision measured in microarcseconds—millions of times better than Eddington's optical measurements. These ultra-precise observations have confirmed Einstein's light-bending predictions to accuracy levels of 0.02 percent, pushing the classical tests far beyond what their original designers could have imagined.

The gravitational redshift—Einstein's prediction that light climbing out of gravitational wells would lose energy and appear redshifted—proved even more challenging to measure precisely. Early attempts in 1925 to observe redshift in the light from white dwarf stars were suggestive but not conclusive. The definitive terrestrial confirmation came in 1959 with the Pound-Rebka experiment at Harvard University, which used the extraordinarily narrow spectral lines produced by the Mössbauer effect to measure gravitational redshift over just 22.5 meters of elevation change.

The Pound-Rebka experiment achieved precision of about 1 percent in confirming Einstein's predictions, but even this remarkable accuracy has been surpassed by modern atomic clock experiments that can detect gravitational time dilation over elevation changes as small as 33 centimeters.

The Precision Revolution

Modern tests of relativity have entered an era of precision that would have seemed impossible to Einstein's generation. Atomic clocks now achieve stability of one part in 10^{18}—meaning they would gain or lose less than one second over the entire age of the universe. This extraordinary precision has transformed gravitational time dilation from a barely detectable effect into a phenomenon that can be measured with routine accuracy in everyday laboratory environments.

In 2010, researchers at the National Institute of Standards and Technology demonstrated gravitational time dilation by simply raising an optical atomic clock by one-third of a meter on their laboratory bench. The elevated clock gained time at exactly the rate predicted by Einstein's equations, confirming general relativity with precision that exceeded the original Pound-Rebka experiment by orders of magnitude.

These ultra-precise clocks have revealed that we live in a world where gravitational time dilation affects every moment of our existence. Every time we climb stairs, ride elevators, or even stand up from sitting, we move between regions where time flows at measurably different rates. The effects are tiny by human standards, but they represent real differences in the passage of time itself—differences that modern atomic clocks can detect and measure with extraordinary accuracy.

The precision revolution extends beyond terrestrial experiments to astronomical observations. Binary pulsars—pairs of neutron stars orbiting each other—provide natural laboratories for testing relativity in gravitational fields far stronger than anything in our solar system. The Hulse-Taylor pulsar, discovered in 1974, has been monitored for decades with timing precision of microseconds, revealing the gradual orbital decay predicted by Einstein's theory as the system loses energy through gravitational wave emission.

Radio Wave Time Delays and Spacecraft Tests

One of the most elegant confirmations of Einstein's theory came from radio experiments conducted with NASA's Mariner spacecraft in the 1970s. As these spacecrafts traveled beyond Mars, scientists could send radio signals from Earth, have them reflected by the spacecraft, and measure the round-trip travel time with extraordinary precision.

When the radio signals passed close to the Sun—traveling through regions of curved spacetime—they experienced delays exactly as Einstein's theory predicted. The Sun's gravitational field doesn't just bend the path of radio waves; it also slows their propagation through curved spacetime, creating measurable delays in communication with distant spacecraft.

Dr. John Anderson, who led these experiments at the Jet Propulsion Laboratory, found that the measured delays matched Einstein's predictions to within 2-4 percent accuracy. For Mariner VI, the maximum measured delay was about 204 microseconds, compared to Einstein's predicted value of 200 microseconds. This tiny discrepancy—just 4 microseconds—was within the experimental uncertainty and provided strong confirmation that alternative theories of gravity proposed by other physicists were inconsistent with observations.

These spacecraft tests represented a fourth type of relativity experiment, proposed as early as 1964, that complemented the three classical tests by probing the effects of curved spacetime on electromagnetic wave propagation over vast distances. The experiments demonstrated that Einstein's insights about space and time apply not just to local phenomena but to the structure of the solar system itself.

Binary Pulsars: Natural Relativity Laboratories

The discovery of binary pulsars in the 1970s opened an entirely new frontier for testing Einstein's theory under extreme conditions. These systems consist of two neutron stars—objects so dense that they compress the mass of the Sun into spheres only 20 kilometers in diameter—orbiting each other with periods ranging from minutes to hours.

The neutron star that we observe as a pulsar serves as an incredibly precise cosmic clock, emitting regular pulses of radio waves with the accuracy of the finest atomic timepieces. But unlike artificial clocks, pulsars operate in gravitational environments billions of times stronger than Earth's gravity, where relativistic effects become dramatically amplified.

Binary pulsar systems exhibit multiple relativistic effects simultaneously. The orbital periods decay over time as the system loses energy through gravitational wave emission, exactly as Einstein's theory predicts. The orbits precess due to relativistic effects on orbital motion. The pulse arrival times vary due to time dilation effects as the pulsar moves through varying gravitational fields in its elliptical orbit.

The precision of pulsar timing has reached extraordinary levels. The pulse arrival times can be measured to accuracy of microseconds or better, allowing astronomers to detect orbital changes that accumulate over years or decades. Some binary pulsars have been monitored for more than 40 years, providing tests of Einstein's theory with precision that exceeds laboratory experiments by many orders of magnitude.

Gravitational Wave Astronomy

The February 2016 announcement of gravitational wave detection by LIGO represented the culmination of decades of technological development aimed at testing Einstein's most subtle prediction. Gravitational waves—ripples in spacetime itself—had been predicted by Einstein in 1916 but were considered too weak to ever detect directly.

The LIGO detection required pushing measurement precision to almost incomprehensible limits. The interferometers must detect changes in the length of their 4-kilometer arms that are smaller than 1/10,000th the width of a proton. To achieve this sensitivity, LIGO must isolate its measurements from countless sources of noise: seismic vibrations, thermal fluctuations, quantum uncertainty, and even the gravitational effects of traffic on nearby highways.

The first detected gravitational wave signal, designated GW150914, came from the merger of two black holes approximately 36 and 29 times the mass of our Sun, located about 1.3 billion light-years from Earth. The signal lasted only about 0.2 seconds as observed by LIGO, but it contained information about gravitational physics in the most extreme environment ever studied.

During the final moments of the merger, the black holes were orbiting each other at velocities approaching half the speed of light, in gravitational fields so intense that they curved spacetime nearly to its breaking point. The time dilation effects in such an environment would be extreme beyond imagination—if conscious observers could somehow survive near such a merger, they would experience time flowing at a dramatically different rate than distant observers watching the event unfold.

The gravitational wave signal itself provided an exquisite test of Einstein's field equations in the strong-field regime where previous tests had been impossible. The detected waveform matched theoretical predictions derived from Einstein's equations with remarkable precision, confirming that general relativity accurately describes spacetime behavior even under the most extreme conditions in the universe.

Precision Tests with Atomic Fountains

The development of optical atomic clocks has pushed gravitational time dilation measurements to new levels of precision that probe the foundations of Einstein's equivalence principle. These devices use laser light to probe atomic transitions with frequencies measured to 18 decimal places, creating clocks so stable that they can detect gravitational time dilation effects that were previously considered unmeasurable.

Optical atomic clocks have confirmed Einstein's predictions in laboratory settings with precision that exceeds one part in 10^{16}. They can detect the gravitational time dilation difference between their own height and the height of a second clock placed just one meter higher or lower. This sensitivity has opened new possibilities for using atomic clocks as gravitational sensors, capable of detecting underground mass distributions or monitoring changes in Earth's gravitational field.

These ultra-precise measurements have also enabled tests of the universality of gravitational time dilation—the prediction that all types of clocks, regardless of their internal mechanisms, should show identical time dilation effects when placed in the same gravitational environment. By comparing different types of atomic clocks simultaneously, physicists have confirmed this universality to extraordinary precision, ruling out alternative theories of gravity that predict slight differences in how different atomic species respond to gravitational fields.

Testing the Equivalence Principle

The equivalence principle—Einstein's insight that gravitational and inertial mass are identical—continues to be tested with ever-increasing precision using sophisticated torsion balance experiments. These tests compare the acceleration of different materials in Earth's gravitational field, searching for tiny violations of the principle that would indicate new physics beyond Einstein's theory.

Modern equivalence principle tests have achieved precision of better than one part in 10^{13}, confirming that different materials fall with identical acceleration to within this extraordinary accuracy. These experiments test not just ordinary matter but also antimatter, different isotopes, and materials with different ratios of binding energy to total mass.

The search for equivalence principle violations is motivated by theoretical predictions from string theory and other approaches to quantum gravity, which suggest that violations might occur at levels just beyond current experimental sensitivity. Each improvement in experimental precision either discovers new physics or constrains theoretical models more tightly.

Frame Dragging and the Gravity Probe B Mission

One of Einstein's most subtle predictions—that rotating massive objects should "drag" spacetime around with them as they spin—was finally confirmed by NASA's Gravity Probe B mission launched in 2004. The mission used four ultra-precise gyroscopes in orbit around Earth to measure how Earth's rotation affects the geometry of nearby spacetime.

The gyroscopes were perfect spheres made of fused quartz, polished to tolerances that made them among the most perfectly round objects ever created. As they orbited Earth, the orientation of their spin axes was affected by two relativistic effects: geodetic precession due to Earth's gravitational field, and frame dragging due to Earth's rotation.

After years of careful data analysis to separate the tiny relativistic signals from various sources of systematic error, Gravity Probe B confirmed both effects with precision of about 19 percent for geodetic precession and 19 percent for frame dragging. While these precisions were lower than initially hoped due to unexpected complications, they provided the first direct measurement of frame dragging around Earth.

The Laser Ranging Revolution

Lunar laser ranging—the practice of bouncing laser pulses off retroreflectors placed on the Moon by Apollo astronauts and Soviet lunar missions—has provided some of the most precise tests of Einstein's theory over astronomical distances. By measuring the round-trip travel time of laser pulses with precision of a few centimeters, scientists can monitor the Moon's orbit with extraordinary accuracy.

These measurements have confirmed multiple predictions of general relativity, including the rate of precession of the Moon's orbit, the effects of the Sun's gravity on Earth-Moon dynamics, and the constancy of the gravitational constant over time. The precision achieved through decades of lunar ranging has ruled out many alternative theories of gravity and confirmed Einstein's predictions with accuracy approaching one part in 10^{11}.

Pulsar Timing Arrays and Nanohertz Gravitational Waves

The most recent frontier in testing Einstein's theory involves using networks of pulsars across our galaxy as a detector for very low-frequency gravitational waves. Pulsar timing arrays monitor dozens of millisecond pulsars simultaneously, searching for correlated changes in pulse arrival times that would indicate gravitational waves passing through our region of the galaxy.

These experiments are sensitive to gravitational waves with periods of years to decades—frequencies far below what ground-based detectors like LIGO can observe. Such waves would be produced by supermassive black hole mergers in distant galaxies, creating a gravitational wave background that should be detectable through its subtle effects on pulsar timing.

Recent results from pulsar timing arrays have provided evidence for this gravitational wave background, opening a new window for testing Einstein's theory in previously unexplored regimes. The detection methods require tracking pulsar timing with precision of tens of nanoseconds over periods of decades, representing one of the most demanding precision experiments ever attempted.

Looking Toward Future Tests

As experimental precision continues to improve, future tests of Einstein's theory will probe even more extreme regimes and subtle effects. Space-based gravitational wave detectors like LISA will observe mergers of supermassive black holes, providing tests of general relativity in gravitational fields even stronger than those accessible to ground-based detectors.

Next-generation atomic clocks promise precision approaching one part in 10^{19}, enabling tests of gravitational time dilation that could detect effects from quantum gravity theories that predict tiny violations of Einstein's equivalence principle. Ultra-stable optical clocks may eventually become sensitive enough to detect gravitational waves directly through the time dilation effects they create.

Advances in quantum sensing may also enable new types of relativity tests. Quantum gravimeters and quantum inertial sensors could provide unprecedented sensitivity to gravitational effects, while quantum entanglement experiments may probe how relativity affects quantum mechanical systems in new ways.

The Enduring Power of Einstein's Vision

As Dr. Patel reflected on the LIGO detection and its implications, he was struck by the remarkable fact that Einstein's century-old insights continue to pass every experimental test with flying colors. From the gentle time dilation effects measured by atomic clocks in earthbound laboratories to the extreme spacetime curvature probed by gravitational wave detectors, Einstein's geometric description of gravity has proven accurate in every regime where it can be tested.

The precision of modern tests exceeds anything Einstein could have imagined when he first formulated his theory. Yet his mathematical framework continues to predict experimental results with extraordinary accuracy, suggesting that our understanding of spacetime geometry captures something fundamental about the structure of physical reality.

Perhaps most remarkably, each new test of Einstein's theory has revealed new phenomena rather than contradicting existing predictions. Gravitational waves, binary pulsar evolution, frame dragging effects, and cosmological time dilation have all been discovered exactly as Einstein's equations predicted, opening new frontiers for astrophysics and cosmology rather than requiring modifications to the underlying theory.

The story of testing Einstein's limits demonstrates the power of mathematical physics to describe natural phenomena across an enormous range of scales and conditions. From atomic clocks measuring time dilation over centimeters to gravitational wave detectors revealing black hole mergers billions of light-years away, Einstein's insights about the nature of space and time provide a unified framework that encompasses the entire observable universe.

As experimental technology continues to advance, future tests will undoubtedly probe even more extreme regimes and achieve even greater precision. But if the history of relativity testing is any guide, these experiments will likely reveal new wonders about the universe while continuing to confirm the fundamental correctness of Einstein's revolutionary insights about space, time, and gravity.

The limits of Einstein's theory, if they exist, remain beyond our current ability to detect them. Each new generation of experiments pushes deeper into the foundations of spacetime, seeking the boundaries where general relativity might break down and new physics emerges. But so far, the universe continues to operate exactly as Einstein envisioned—a cosmos where space and time form a unified, dynamic fabric that responds to matter and energy in ways both beautiful and mathematically precise.

In testing the limits of Einstein's theory, we are ultimately testing the limits of our understanding of reality itself. The extraordinary precision of modern experiments reveals a universe that operates according to mathematical principles of remarkable elegance and universality. As we continue to push the boundaries of measurement and observation, we carry forward Einstein's fundamental insight that the deepest truths about nature can be discovered through human reason, expressed in mathematical form, and tested against the observable universe with ever-increasing precision and wonder.

Chapter 15: Future Implications

Professor Yuki Tanaka stood at the podium of the International Space Development Conference in Tokyo, her presentation slides showing a sleek spacecraft design unlike anything humanity had ever attempted to build. The vessel stretched nearly a kilometer in length, its fusion engines capable of accelerating it to 20% of light speed—fast enough to reach Proxima Centauri, our nearest stellar neighbor, in just over 20 years. But as she explained to the assembled engineers, physicists, and policymakers, the revolutionary propulsion system was only half the challenge.

The other half was Einstein.

"When this ship returns to Earth after a 40-year round trip," Yuki explained, advancing to her next slide showing two side-by-side calendars, "the crew will have aged only about 38 years, while 40 years will have passed on Earth. The first interstellar explorers won't just be traveling through space—they'll be traveling through time itself, returning to find that everyone they knew has aged two years more than they have."

The audience sat in contemplative silence as the implications sank in. This wasn't science fiction anymore—it was engineering reality waiting for technology to catch up with physics. Einstein's insights about time dilation, proven countless times in laboratory experiments and astronomical observations, would soon become practical considerations for humanity's expansion beyond the solar system.

This is the future that Einstein's universe promises: a cosmos where time itself becomes malleable through technology, where the effects we've explored throughout this book transform from curious physical phenomena into the fundamental constraints and opportunities that will shape human civilization's next chapter among the stars.

The New Age of Space Exploration

As humanity prepares for its next great leap into space, Einstein's theories of relativity are transitioning from scientific curiosities to engineering requirements. Every mission beyond Earth's immediate vicinity must now account for relativistic effects that previous generations of space explorers could safely ignore.

Current Mars missions already encounter measurable time dilation effects, though they remain negligible for practical purposes. A spacecraft traveling to Mars at conventional velocities experiences time dilation of only a few microseconds over the entire journey. But as propulsion technology advances and mission distances increase, these effects will become significant factors in mission planning.

The proposed Breakthrough Starshot initiative, which aims to send tiny probes to nearby star systems at 20% of light speed, represents humanity's first serious encounter with relativistic engineering challenges. At such velocities, time dilation factors of about 2% become significant for mission operations, communication timing, and data transmission protocols.

For crewed missions to nearby stars, the implications become profound. A journey to Proxima Centauri at 10% of light speed would take about 44 years from Earth's perspective, but only about 43.8 years from the crew's perspective—a difference of several months. At 50% of light speed, the journey would take about 9.3 years for Earth observers but only about 8 years for the travelers.

These time differences aren't just technical curiosities—they represent fundamental challenges for maintaining contact with Earth, coordinating mission activities, and planning for the psychological and social implications of relativistic space travel. Mission planners must now consider not just the physics of propulsion and life support, but the physics of time itself.

The Technology of Relativistic Flight

The engineering challenges of achieving relativistic velocities push the boundaries of known technology in ways that reveal new applications for Einstein's insights. Fusion rockets, antimatter engines, and even more exotic propulsion concepts like solar sails and ramjet designs all must account for relativistic effects as spacecraft approach significant fractions of light speed.

Consider the relativistic rocket equation, which extends classical rocket physics to include Einstein's mass-energy relationship. As a spacecraft accelerates to high velocities, the effective mass of its fuel increases, requiring exponentially more energy to achieve each additional increment of velocity. This means that reaching even 10% of light speed requires fuel mass ratios that challenge conventional spacecraft design.

Ion drives and other electric propulsion systems offer more promising approaches to relativistic flight, but they require enormous amounts of electrical power. Solar panels become less effective as spacecraft travel farther from the Sun, while nuclear reactors must be designed to operate reliably for decades or centuries. The engineering challenges multiply when we consider that relativistic effects alter the behavior of all onboard systems.

Atomic clocks aboard relativistic spacecraft will run at different rates than Earth-based clocks, affecting navigation, communication, and experimental timing. Computer processors will experience time dilation that affects their apparent processing speeds when compared to Earth-based systems. Even biological processes will be influenced by the altered flow of time, requiring new approaches to life support and medical monitoring.

The development of relativistic propulsion technology may also reveal new physics beyond Einstein's theories. Some proposed propulsion concepts, such as Alcubierre drives that manipulate spacetime geometry itself, would require exotic forms of matter with properties that push the boundaries of known physics. While such technologies remain highly speculative, they illustrate how the pursuit of relativistic flight continues to drive fundamental physics research.

Communication Across Time

One of the most practical challenges facing future space exploration involves maintaining communication with relativistic spacecraft. Radio signals traveling between Earth and a fast-moving spacecraft experience Doppler shifts that affect both frequency and timing. These effects become more pronounced as spacecraft velocities increase, requiring new approaches to interstellar communication.

The time delay for round-trip communication with nearby stars is already measured in years—about 8.6 years for a signal to travel to Proxima Centauri and back. But relativistic effects add complexity beyond simple light-travel delays. Signals from a spacecraft experiencing significant time dilation will appear to be transmitted at different rates than signals from Earth-based transmitters.

Future deep space missions will need communication protocols that account for these relativistic effects. Ground controllers must understand that responses to their commands will arrive not just delayed by distance, but modified by the time dilation experienced by the spacecraft. Mission timelines must be planned around the fact that the passage of time itself becomes a variable factor in operation scheduling.

This challenge extends to navigation and course corrections. GPS-like navigation systems for interstellar spacecraft must account for relativistic effects on signal propagation and timing. The precision required for interstellar navigation is orders of magnitude greater than anything needed for solar system exploration, demanding new applications of Einstein's theories to spacecraft guidance and control.

Biological and Psychological Implications

The biological implications of relativistic space travel represent an entirely new frontier in space medicine. While astronauts on conventional space missions experience microgravity and radiation exposure, future interstellar travelers will also contend with the biological effects of time dilation itself.

From a purely physical perspective, biological processes should continue normally from the travelers' reference frame. Aging, metabolism, and cellular repair mechanisms will proceed at their usual rates as measured by onboard clocks. However, the psychological implications of knowing that time is passing differently may create unprecedented challenges for crew mental health.

Imagine being part of a crew that departs Earth knowing that everyone you've ever known will age faster than you during your journey. Friends and family members will experience years of life while you experience months. By the time you return, entire generations may have passed on Earth while you remain relatively young.

This temporal displacement could create profound psychological stress that goes beyond anything experienced by previous explorers. Traditional space missions maintain real-time communication with Earth, allowing crews to remain connected with their home communities. Relativistic missions will experience increasingly delayed and time-dilated communication, creating a form of temporal isolation unlike anything in human experience.

Space agencies are beginning to study these psychological implications, developing new approaches to crew selection, training, and support that account for the unique stresses of relativistic travel. Virtual reality systems might help maintain connection with Earth communities, while advanced AI systems could provide companionship and psychological support during the long journey through both space and time.

Technological Spin-offs and Applications

The pursuit of relativistic space technology is already driving innovations that benefit terrestrial applications. Advanced atomic clocks developed for precision spacecraft navigation are finding applications in financial markets, telecommunications, and scientific research. Fusion propulsion research contributes to the development of clean energy technology on Earth.

Perhaps more significantly, the engineering challenges of relativistic flight are advancing our understanding of materials science, energy storage, and system reliability. Spacecraft that must operate for decades while experiencing extreme accelerations and radiation environments require materials and components that push the boundaries of engineering capability.

These advances often find unexpected applications in other fields. Research into long-term spacecraft autonomy contributes to the development of artificial intelligence and robotic systems. Studies of closed-loop life support systems advance our understanding of ecological balance and sustainable technology. Even the psychological research needed for relativistic crews provides insights into human adaptation and resilience.

The economic implications extend beyond direct technological applications. The development of relativistic propulsion technology could enable asteroid mining operations that access resources throughout the solar system and beyond. The ability to travel at significant fractions of light speed would make previously unreachable asteroids and comets accessible within reasonable mission timescales.

Quantum Mechanics and Relativistic Technologies

As we push toward relativistic velocities, the intersection between Einstein's relativity and quantum mechanics becomes increasingly important for practical technology. Quantum effects that are negligible at everyday velocities become significant when particles and systems approach light speed.

Quantum computers designed to operate aboard relativistic spacecraft must account for how high velocities affect quantum states and entanglement. The GPS satellites we rely on today already require corrections for both relativistic and quantum effects, but future relativistic missions will encounter these interactions in much more extreme forms.

The development of quantum sensors for relativistic spacecraft navigation represents a particularly promising area of research. Quantum gravimeters and quantum inertial sensors could provide navigation capabilities that remain accurate even when conventional GPS-like systems become impractical due to distance and time delays.

Quantum communication systems might also offer solutions to some challenges of relativistic space travel. Quantum entanglement could potentially provide instantaneous communication links that bypass the light-speed delays inherent in conventional radio communication, though the practical implementation of such systems remains highly challenging.

Medical and Biological Applications

The precision timing and measurement capabilities developed for relativistic physics research continue to find applications in medical technology. MRI machines rely on Einstein's unification of electromagnetic forces, while precision atomic clocks enable medical timing applications that require extraordinary accuracy.

Future medical applications may exploit relativistic effects more directly. Time dilation could theoretically be used to slow biological processes during medical procedures, though the velocities or gravitational fields required to create significant effects remain far beyond current technology.

The study of how biological systems respond to time dilation effects may reveal new insights into aging, metabolism, and cellular repair mechanisms. Understanding how life processes are affected by altered temporal flow could contribute to longevity research and medical treatments that manipulate the rate at which biological processes occur.

Advanced medical imaging systems continue to benefit from relativistic physics research. Particle beam therapy for cancer treatment uses relativistic effects to precisely control the energy deposition of high-speed particles. Future medical technologies may exploit other relativistic phenomena to diagnose and treat diseases in ways we're only beginning to imagine.

The Economics of Relativistic Time

The economic implications of time dilation effects will become increasingly important as human activity expands beyond Earth. Financial markets already rely on GPS timing precision for high-frequency trading, but relativistic effects will create new challenges for interplanetary commerce.

Consider a mining operation on a distant asteroid that uses relativistic spacecraft to transport materials back to Earth. The time dilation experienced during high-speed transport creates complications for commodity pricing, contract fulfillment, and financial planning. How do you price materials that take years to transport when the transport time is different for the cargo than for Earth-based markets?

Insurance companies will need to develop new models for covering relativistic missions where the passage of time itself becomes a variable risk factor. Legal systems must evolve to handle contracts and agreements between parties experiencing different rates of temporal flow.

The development of interstellar commerce will require new economic frameworks that account for the fact that business transactions might involve parties whose clocks are running at different rates. The concept of "simultaneous" contracts becomes meaningless when different parties disagree about what constitutes the same moment in time.

Educational and Cultural Evolution

As relativistic effects become part of everyday human experience through advanced space travel, educational systems will need to prepare people for life in Einstein's universe. Children growing up in a relativistic age will need to understand concepts of time dilation, spacetime geometry, and causal relationships that previous generations could treat as abstract theoretical concepts.

The cultural implications are equally profound. How do societies maintain continuity when some members are traveling at relativistic speeds and aging at different rates? What happens to family structures when parents and children might age at different rates depending on their travel patterns?

Literature, art, and entertainment are already beginning to explore these themes as relativistic travel transitions from science fiction to engineering reality. The movies *Interstellar* showcased how Hollywood is grappling with the dramatic possibilities of time dilation effects, while science fiction literature increasingly incorporates accurate relativistic physics into its narratives.

Future educational curricula will likely include practical training in relativistic navigation, time dilation calculations, and the social protocols needed for societies spanning multiple reference frames. The mathematical literacy required for citizens of a relativistic civilization will be significantly higher than what previous generations needed.

The Search for New Physics

While Einstein's theories continue to pass every experimental test, the pursuit of relativistic technology drives the search for physics beyond relativity. The extreme conditions required for significant relativistic effects often reveal the limits where Einstein's theories might break down and new physics emerges.

Experiments with ultra-high precision atomic clocks are beginning to probe the intersection between general relativity and quantum mechanics, searching for effects that neither theory alone can explain. The development of gravitational wave detectors has opened new windows into the universe that may reveal phenomena beyond Einstein's predictions.

The engineering challenges of achieving significant relativistic velocities may require new physics to solve. Breakthrough propulsion concepts often involve exotic forms of matter or energy that don't exist in Einstein's universe as currently understood. The pursuit of these technologies drives fundamental physics research that could revolutionize our understanding of space, time, and matter.

Future particle accelerators designed to push the boundaries of high-energy physics will encounter relativistic effects in increasingly extreme forms. These facilities may reveal new particles or forces that require extensions to Einstein's theories, opening new frontiers in both fundamental physics and practical technology.

Environmental and Sustainability Considerations

The development of relativistic technology also raises important questions about environmental impact and resource sustainability. The energy requirements for accelerating massive spacecraft to significant fractions of light speed are enormous, potentially requiring new approaches to energy generation and consumption.

Fusion propulsion systems offer the promise of clean, abundant energy for relativistic flight, but they also require materials and technologies that must be developed sustainably. The mining operations needed to extract fuel and construction materials for relativistic spacecraft could have significant environmental impacts unless carefully managed.

The long-term implications of human expansion to other star systems also raise questions about our responsibility to preserve cosmic environments. How do we ensure that relativistic exploration enhances rather than diminishes the cosmic heritage we're exploring?

These considerations are driving research into sustainable space technology and responsible exploration protocols that consider the long-term implications of human activity across interstellar distances and relativistic timescales.

The Philosophy of Relativistic Existence

Perhaps the most profound implications of living in Einstein's universe involve fundamental questions about the nature of existence, consciousness, and reality itself. If time is relative and malleable, what does this mean for our understanding of life, death, and the meaning of existence?

Relativistic travel raises questions about personal identity and continuity of self. If you embark on a relativistic journey and return to Earth younger than people who stayed behind, are you the same person who departed? How do concepts of aging, maturity, and life experience change when the passage of time itself becomes variable?

These philosophical questions are not merely abstract—they will have practical implications for legal systems, social structures, and personal relationships as relativistic travel becomes commonplace. Societies will need to develop new frameworks for understanding identity, responsibility, and social connections across different temporal reference frames.

The implications extend to our understanding of consciousness itself. If time dilation affects all physical processes, does it also affect the subjective experience of consciousness? Do thoughts, emotions, and awareness unfold more slowly for relativistic travelers, or does consciousness somehow remain immune to temporal effects?

Looking to the Deep Future

As Professor Tanaka concluded her presentation, she displayed a final image: a timeline stretching from the present to thousands of years in the future, showing humanity's gradual expansion across the galaxy through increasingly sophisticated relativistic technology. The early interstellar missions she had described were just the beginning of a transformation that would fundamentally alter the human experience.

In the centuries to come, time dilation might become as commonplace as jet lag is today. Families might routinely include members aging at different rates due to their different travel patterns. Commerce and communication networks will span distances where information takes years to travel and time itself flows at different rates in different locations.

The ultimate implications of Einstein's universe extend beyond even these dramatic changes. As humanity develops technology that approaches the theoretical limits of relativistic physics, we may discover that space and time themselves can be manipulated in ways that Einstein never imagined.

The future implications of time dilation extend far beyond the practical challenges of space travel and technology. They touch on the most fundamental questions about the nature of reality, consciousness, and human existence. As we learn to live in Einstein's universe, we're not just adapting to new technology—we're evolving into a species for whom the relativity of space and time becomes an integral part of the human experience.

The clockmaker's dilemma that we encountered in our first chapter has been resolved, but the solution has opened doors to possibilities that no previous generation could have imagined. We now understand that time is not the absolute, universal backdrop that Newton envisioned, but rather a dynamic aspect of reality that responds to matter, energy, and motion in ways that human technology can harness and exploit.

As we stand on the threshold of the relativistic age, we carry with us Einstein's profound insight that the universe is both stranger and more wonderful than our everyday experience suggests, yet operates according to mathematical principles that human reason can discover and understand. The future belongs to those who can navigate not just through space, but through the malleable dimensions of spacetime itself, using Einstein's insights as their compass for the journey through both the cosmos and time.

The story of time dilation that began with thought experiments about falling elevators and speeding trains now points toward a future where humanity spreads across the galaxy, armed with the knowledge that time itself is one of the universe's most remarkable and manipulable resources. In learning to bend time to our purposes, we fulfill Einstein's ultimate promise: that understanding the deepest truths about the universe empowers us to transcend the limitations of our origins and become something greater than we ever dared to imagine.

PART VI:

THE DEEPER PICTURE

Chapter 16: Quantum Gravity and Time

Dr. Sarah Chen stared at the equations covering every whiteboard in her office at the Perimeter Institute for Theoretical Physics, each symbol representing humanity's most ambitious attempt to understand the deepest nature of reality. For three years, she had been working on one of the most profound puzzles in physics: what happens to time itself when Einstein's smooth, continuous spacetime meets the choppy, quantized world of quantum mechanics.

The equations told a story that challenged everything we thought we knew about the flow of time. According to quantum gravity theories, the spacetime that Einstein described as a smooth, flexible fabric might actually be made of discrete, granular pieces at the smallest possible scales—tiny building blocks of space and time so small that they defied imagination. At the Planck scale, where distances shrink to 10^{-35} meters and time intervals compress to 10^{-43} seconds, the very concepts of "before" and "after" might lose their meaning entirely.

Sarah's research suggested something even more radical: that time dilation, the phenomenon we've explored throughout this book, might not be the whole story. In the quantum realm, time itself might not flow continuously but instead tick forward in discrete quantum jumps, like frames in a cosmic movie reel. If true, this would mean that all the time dilation effects we've observed—from GPS satellites to black holes—are merely the large-scale average of countless quantum time events occurring at scales so small that they create the illusion of smooth temporal flow.

This is the frontier where Einstein's insights about curved spacetime meet the strange world of quantum mechanics, creating a new understanding of time that is more mysterious and more wonderful than either theory alone could reveal.

The Fundamental Incompatibility

To understand why quantum gravity represents such a revolutionary challenge to our understanding of time, we must first appreciate the fundamental incompatibility between Einstein's general relativity and quantum mechanics—two theories that have never failed an experimental test, yet cannot coexist in the same mathematical framework.

Einstein's general relativity describes gravity not as a force but as the curvature of a smooth, continuous spacetime. Time flows smoothly in this picture, capable of being stretched and compressed by gravitational fields, but always maintaining its continuous character. The time dilation effects we've explored throughout this book emerge naturally from this smooth geometry—clocks run slower in stronger gravitational fields because time itself flows more slowly in regions where spacetime is more severely curved.

Quantum mechanics, however, operates according to entirely different principles. The three other fundamental forces—electromagnetism, the weak nuclear force, and the strong nuclear force—are all described by quantum field theories where forces are transmitted by discrete particles. The electromagnetic force, for example, is carried by photons, while the weak nuclear force is mediated by W and Z bosons.

This quantized nature of forces suggests that gravity should also be quantized, transmitted by hypothetical particles called gravitons. But when physicists attempt to apply quantum mechanical principles to gravity, the mathematics becomes impossible, producing infinite answers where finite numbers should appear. The smooth, continuous spacetime of general relativity seems fundamentally incompatible with the discrete, quantized nature of quantum mechanics.

This incompatibility becomes acute when we consider extreme environments where both quantum effects and gravitational effects are important—near black holes, in the early universe, or at the Planck scale where quantum fluctuations of spacetime itself are expected to become significant.

The Granular Universe

One of the most promising approaches to quantum gravity, called loop quantum gravity, makes a radical proposal: that spacetime itself has a granular structure at the smallest scales. Just as matter is made of discrete atoms despite appearing continuous in everyday experience, spacetime might be composed of fundamental units of space and time that create the illusion of continuity at larger scales.

In loop quantum gravity, space is not infinitely divisible but instead has a minimum possible size—the Planck length of about 10^{-35} meters. Similarly, time cannot be divided into arbitrarily small intervals but has fundamental units of Planck time, roughly 10^{-43} seconds. Below these scales, the very concepts of space and time lose their meaning.

This granular structure would fundamentally alter our understanding of time dilation. Instead of time flowing smoothly and continuously, it would tick forward in discrete quantum steps. The time dilation we observe in gravitational fields would emerge from the statistical average of countless quantum temporal events, each occurring at the Planck scale.

Imagine time as being composed of individual "chronons"—fundamental units of temporal existence that cannot be subdivided further. Time dilation would occur because gravitational fields affect how these chronons are distributed and how frequently they occur. In stronger gravitational fields, chronons might be spaced further apart in some fundamental sense, causing time to appear to flow more slowly when averaged over many chronons.

This picture suggests that the smooth time dilation effects described by Einstein's equations are actually emergent phenomena—large-scale averages of underlying quantum processes that operate according to very different rules. The continuous stretching of time that we observe near black holes or in accelerating reference frames would be like the apparently smooth flow of water, which is actually composed of countless individual molecules moving according to quantum mechanical principles.

Spin Networks and the Quantum Structure of Spacetime

Loop quantum gravity describes the quantum state of spacetime using mathematical structures called spin networks—abstract graphs where nodes and edges represent the quantum properties of space itself. These networks don't exist in spacetime; rather, they are spacetime, encoding all the geometric information about distances, areas, and volumes in a fundamentally quantum mechanical way.

In this framework, time emerges from the evolution of these spin networks as they change from one configuration to another in discrete steps called spinfoams. Each transition represents a quantum event in the structure of spacetime itself, and the accumulation of many such transitions creates the illusion of smooth temporal flow.

The implications for time dilation are profound. Instead of time being stretched by the continuous curvature of spacetime, temporal effects would arise from changes in how spin networks evolve in different gravitational environments. Stronger gravitational fields might alter the probability distributions governing spin network transitions, causing time to appear to slow down when viewed from regions of weaker gravity.

This quantum mechanical picture of time suggests that all the relativistic effects we've explored—from the GPS satellites correcting for nanosecond timing differences to the extreme time dilation near black holes—are actually manifestations of deeper quantum processes occurring at the Planck scale. The mathematical precision of Einstein's equations would emerge as statistical averages of quantum gravitational phenomena, much as the laws of thermodynamics emerge from the statistical behavior of countless molecules.

The Problem of Time in Quantum Gravity

One of the most puzzling aspects of quantum gravity theories is what physicists call "the problem of time". When Einstein's field equations are quantized using the mathematical techniques of quantum mechanics, something strange happens: time disappears from the fundamental equations entirely.

The Wheeler-DeWitt equation, which represents the quantum mechanical version of Einstein's field equations, contains no explicit time parameter. This suggests that at the most fundamental level, the universe might be static and timeless, with time emerging only as an approximate concept that applies at larger scales.

This disappearance of time in quantum gravity has profound implications for our understanding of temporal flow and time dilation. If time is not fundamental but instead emerges from more basic quantum processes, then the time dilation effects we observe might be secondary phenomena that arise when we average over underlying quantum dynamics that don't themselves involve time in any conventional sense.

Some physicists, following the work of Julian Barbour, have suggested that time itself might be nothing more than a measure of change and motion—a way of parameterizing how configurations of matter and energy evolve, rather than a fundamental aspect of reality. In this view, time dilation would represent different ways of measuring change in different reference frames, rather than actual differences in the flow of some fundamental temporal substance.

Quantum Fluctuations and Spacetime Foam

At the Planck scale, quantum mechanics predicts that spacetime itself should undergo violent fluctuations, creating what physicists call "spacetime foam"—a seething sea of quantum uncertainty where the normal concepts of distance and duration break down entirely. In these quantum fluctuations, virtual particles pop in and out of existence, while the geometry of spacetime itself fluctuates wildly.

These quantum fluctuations of spacetime could have important consequences for time dilation effects. The smooth, predictable time dilation described by Einstein's equations might represent only the average behavior of spacetime over many quantum fluctuations. At sufficiently small scales or under sufficiently extreme conditions, the underlying quantum nature of spacetime might become apparent, causing deviations from Einstein's predictions.

Near black holes, where gravitational fields approach their maximum possible strength, quantum fluctuations of spacetime might become so intense that they alter the nature of time dilation itself. The event horizon—the boundary beyond which time dilation becomes infinite according to classical general relativity—might be modified by quantum effects in ways that fundamentally change our understanding of black hole physics.

Some theories suggest that quantum effects near black holes might prevent the formation of true event horizons, replacing them with quantum barriers that allow information to eventually escape. If true, this would mean that the extreme time dilation effects we expect near black holes are modified by quantum mechanics in ways that preserve the flow of information and prevent the paradoxes that arise when general relativity is pushed to its limits.

Emergent Time and the Arrow of Temporal Flow

Quantum gravity theories must also address one of the deepest mysteries about time: why does it have a direction? The fundamental laws of physics are largely symmetric with respect to time—they work equally well whether time flows forward or backward. Yet we experience time as having a clear direction, flowing from past to future with no possibility of reversal.

In quantum gravity frameworks where time emerges from more fundamental quantum processes, the arrow of time might arise from the statistical properties of quantum spacetime evolution. Just as the second law of thermodynamics emerges from the statistical behavior of countless particles, the directional flow of time might emerge from the quantum dynamics of spacetime itself.

This emergent arrow of time could have important implications for time dilation. The effects we observe—clocks running slower in gravitational fields, time passing differently for rapidly moving observers—might be manifestations of deeper asymmetries in the quantum structure of spacetime that create the directional flow of temporal experience.

Some theories suggest that the expansion of the universe provides the fundamental source of time's arrow, creating a preferred direction in the quantum evolution of spacetime that manifests as the forward flow of time. In this picture, time dilation effects would be intimately connected to cosmological evolution, linking the local stretching of time to the global expansion of the universe.

String Theory and Higher Dimensions

String theory, another major approach to quantum gravity, proposes that the fundamental constituents of reality are not point particles but tiny vibrating strings existing in ten or eleven dimensions of spacetime. In string theory, gravity emerges naturally from the quantum mechanical properties of these strings, potentially solving the mathematical inconsistencies that plague other approaches to quantum gravity.

String theory's treatment of time is particularly intriguing. The theory suggests that time itself might have additional dimensions beyond the single temporal dimension we experience.

These extra time dimensions would be "compactified"—rolled up so small that we don't notice them in everyday experience—but they could have important consequences for the nature of time dilation and temporal flow.

In string theory frameworks, time dilation effects might arise from the way gravitational fields affect the geometry of these higher-dimensional spaces. The stretching of time that we observe in gravitational fields could reflect changes in the compactification of extra temporal dimensions, creating effects that appear as time dilation when viewed from our limited three-dimensional spatial and one-dimensional temporal perspective.

String theory also suggests the possibility of exotic objects called "time machines" or "closed time-like curves"—configurations of spacetime where time loops back on itself. While such objects are highly speculative and face numerous theoretical obstacles, they illustrate how quantum gravity theories might predict temporal phenomena that go far beyond the time dilation effects described by classical general relativity.

Experimental Challenges and Future Prospects

Testing quantum gravity theories presents extraordinary experimental challenges because the effects typically become important only at the Planck scale—energies and distances so extreme that they are far beyond the reach of any conceivable technology. The energies required to directly probe quantum gravitational effects exceed those of the most powerful particle accelerators by more than fifteen orders of magnitude.

However, some effects of quantum gravity might be accessible to observation in extreme astrophysical environments. Black holes, neutron stars, and the early universe all involve conditions where quantum gravitational effects could become important, potentially creating observable deviations from Einstein's predictions.

Future gravitational wave detectors might be sensitive enough to detect quantum corrections to Einstein's equations in the signals from black hole mergers. Ultra-precise atomic clocks could potentially detect quantum fluctuations in the flow of time itself. Space-based experiments might be able to measure subtle violations of the equivalence principle that could arise from quantum gravitational effects.

The development of quantum sensing technologies offers particular promise for probing the quantum nature of spacetime. Quantum gravimeters and quantum inertial sensors might eventually become sensitive enough to detect the granular structure of space and time predicted by loop quantum gravity theories.

Implications for Our Understanding of Reality

The quest to understand quantum gravity is forcing physicists to reconsider the most fundamental aspects of reality—not just the nature of space, time, and gravity, but the very concepts of existence, causality, and physical law. If time itself emerges from quantum processes rather than being fundamental, then the time dilation effects we've explored throughout this book might be just the tip of an iceberg of temporal phenomena that we're only beginning to understand.

The possibility that spacetime has a granular structure raises profound questions about the nature of continuity and infinity in physics. If space and time are composed of discrete units, then the smooth mathematics of calculus and differential geometry that underlies Einstein's theories might be approximations to more fundamental discrete mathematical structures.

The implications extend beyond physics to philosophy, consciousness, and the nature of existence itself. If time is not fundamental but emerges from quantum processes, what does this mean for our subjective experience of temporal flow? How do consciousness and free will relate to the quantum structure of spacetime?

These questions touch on some of the deepest problems in science and philosophy, connecting quantum gravity research to fields ranging from neuroscience to computer science to the philosophy of mind. The answers might fundamentally change how we understand not just the physical universe but our place within it.

The Continuing Journey

As Dr. Chen stepped back from her whiteboards covered with equations attempting to merge relativity and quantum mechanics, she reflected on how far the journey from Einstein's initial insights had brought us. The time dilation effects that began as curious predictions from thought experiments about falling elevators had led to a complete reconceptualization of the nature of time itself.

The smooth, continuous time of Einstein's relativity—capable of being stretched and compressed but always flowing uniformly—might give way to a deeper understanding where time emerges from quantum processes occurring at the smallest possible scales. The GPS satellites orbiting overhead, continuously correcting for relativistic effects, might be operating in a universe where time itself ticks forward in discrete quantum steps too small to observe directly but fundamental to the nature of temporal existence.

This quantum mechanical picture of time doesn't diminish the importance of the relativistic effects we've explored throughout this book. Instead, it suggests that these effects are manifestations of even deeper principles governing the structure of reality. The time dilation experienced by high-speed travelers, the gravitational effects on time near massive objects, and the cosmological stretching of time in the expanding universe all emerge as large-scale consequences of quantum processes occurring at the Planck scale.

The quest to understand quantum gravity continues to drive the boundaries of human knowledge, revealing new aspects of space and time that challenge our deepest assumptions about the nature of existence. Whether through loop quantum gravity, string theory, or approaches yet to be discovered, physicists continue to probe the quantum foundations of Einstein's spacetime, seeking to understand how the familiar world of classical physics emerges from the strange quantum realm that underlies all of reality.

As we stand at the frontier of quantum gravity research, we carry with us the profound realization that time—the most fundamental aspect of our existence—may be far more mysterious and wonderful than even Einstein's revolutionary insights suggested. The journey from Newton's absolute time through Einstein's relative time to the quantum time of modern theories reveals the extraordinary power of human reason to penetrate the deepest mysteries of the cosmos while revealing new questions that push the boundaries of knowledge ever further into the unknown.

The story of time dilation has led us from the practical engineering of GPS satellites to the mathematical poetry of quantum gravity, demonstrating that the universe operates according to principles both stranger and more beautiful than everyday experience could ever suggest. In learning to understand time's quantum nature, we continue Einstein's grand project of comprehending the universe through mathematical insight and experimental verification, discovering that reality is always more remarkable than our boldest theoretical speculations dared to imagine.

Chapter 17: The Arrow of Time

Dr. Michael Foster sat in his laboratory at Cambridge University, staring at a cup of coffee that had been sitting on his desk for the past hour. The once-steaming beverage had cooled to room temperature, its heat dispersed into the surrounding air in accordance with the inexorable laws of thermodynamics. As a physicist specializing in statistical mechanics, Michael understood perfectly well why his coffee had cooled—countless molecular collisions had transferred thermal energy from the hot liquid to the cooler environment until equilibrium was reached.

But what fascinated Michael wasn't the cooling itself—it was the stubborn, absolute certainty that his coffee would never spontaneously reheat. He could wait forever, and despite the fact that the fundamental laws governing molecular motion work equally well in both directions of time, those molecules would never spontaneously reorganize themselves to make his coffee hot again. The cream he had stirred in that morning would never spontaneously separate back into distinct swirls of white and brown.

This simple observation touches on one of the deepest mysteries in physics: why does time have a direction at all? Michael's cooling coffee embodies what the British astrophysicist Arthur Eddington called "time's arrow"—the fundamental asymmetry of time that distinguishes past from future and gives temporal flow its relentless forward march. It's a mystery that connects intimately to the time dilation effects we've explored throughout this book, revealing that Einstein's malleable time not only bends and stretches but also flows in a preferred direction that shapes the very structure of physical reality.

The Eddington Insight

The concept of time's arrow emerged from Arthur Eddington's profound meditation on the nature of temporal flow in the 1920s. The same physicist who had confirmed Einstein's prediction about light bending around the Sun during the 1919 solar eclipse was grappling with an even deeper puzzle about the nature of time itself.

In his 1928 book *The Nature of the Physical World,* Eddington articulated the mystery with characteristic clarity: "Let us draw an arrow arbitrarily. If as we follow the arrow we find more and more of the random element in the state of the world, then the arrow is pointing towards the future; if the random element decreases the arrow points towards the past". This connection between temporal direction and increasing disorder—entropy—provided the first scientific criterion for distinguishing past from future.

Eddington's insight was revolutionary because it revealed that time's direction isn't inherent in time itself but emerges from the statistical behavior of matter and energy. Unlike space, which has no preferred direction, time possesses what Eddington called "a singularly interesting property from a philosophical standpoint"—an asymmetry that has no analogue in the three dimensions of space.

This asymmetry creates what we might call the fundamental paradox of temporal direction: the microscopic laws of physics are almost entirely symmetric with respect to time, yet the macroscopic world exhibits a clear preference for one temporal direction over another. Newton's laws, Maxwell's equations for electromagnetism, and even Einstein's field equations for general relativity all work equally well whether time runs forward or backward. Yet eggs don't spontaneously unscramble, rivers don't flow uphill, and coffee doesn't spontaneously reheat.

The Thermodynamic Arrow

The most familiar manifestation of time's arrow appears in thermodynamics through the second law, which states that the entropy of an isolated system always increases or remains constant—it never decreases. This thermodynamic arrow of time provides a statistical explanation for why certain processes appear irreversible despite being built from fundamentally reversible microscopic interactions.

Consider the molecular ballet occurring in Michael's cooling coffee. Each individual collision between a caffeine molecule and a water molecule, or between water molecules and air molecules, follows Newton's laws perfectly. If we could somehow reverse the velocities of every single molecule at any given moment, the entire process would run backward, and the coffee would spontaneously heat up as thermal energy flowed back from the environment.

But this reversal is so astronomically unlikely that it effectively never occurs. The probability of all the randomly moving molecules spontaneously coordinating their motion to transfer heat from the cooler air back into the warmer coffee is infinitesimally small—roughly equivalent to the probability that every air molecule in the room would simultaneously move to one corner, creating a vacuum everywhere else.

The second law of thermodynamics doesn't forbid such spontaneous reversals—it simply recognizes that they are so improbable as to be practically impossible. As British physicist Sir Alfred Brian Pippard noted, "no evidence has ever been presented that the Second Law breaks down under any circumstances". The thermodynamic arrow emerges not from fundamental physical laws but from the statistical behavior of vast numbers of particles.

This statistical foundation of time's arrow reveals something profound about the nature of temporal flow. The direction of time isn't written into the fundamental equations of physics—it emerges from the collective behavior of matter and energy as systems evolve from highly organized states toward increasingly disordered configurations.

Time Dilation and the Arrow

The relationship between time's arrow and the time dilation effects we've explored throughout this book is subtle but profound. Einstein's relativity shows us that time itself is malleable—it can be stretched by gravity or compressed by motion—but it doesn't eliminate time's arrow. Instead, relativistic effects occur within the framework of thermodynamic irreversibility, creating complex interactions between temporal flow and temporal direction.

Consider a thought experiment involving two observers: Alice remains on Earth while Bob travels to a nearby star at relativistic speeds. From Alice's perspective, Bob's journey takes 20 years, but due to time dilation, Bob experiences only 10 years during his high-speed travel. When Bob returns, he has aged 10 years less than Alice.

Now here's the crucial point: throughout Bob's journey, entropy continues to increase in both reference frames, but at different rates as measured by the respective observers. Bob's onboard systems—his life support, his computers, his biological processes—all continue to evolve toward higher entropy states, but this evolution appears to occur more slowly from Alice's perspective due to time dilation.

The arrow of time remains consistent in both reference frames, even though time itself flows at different rates. Bob's coffee would still cool during his journey (though Alice would observe this cooling happening in slow motion), and his biological processes would still age him in the normal direction despite the overall time dilation effect.

This reveals that time's arrow is more fundamental than time's rate. While relativity allows time to be stretched and compressed, it doesn't permit time to flow backward or sideways. The thermodynamic arrow continues to point toward the future in all reference frames, even those experiencing extreme time dilation effects.

The Cosmological Arrow

Eddington identified a second arrow of time in the large-scale structure of the universe itself: the cosmological arrow, defined by the direction in which the universe expands. This cosmological arrow provides an independent way of distinguishing past from future that operates on the largest possible scales.

The expanding universe creates a preferred temporal direction that's written into the geometry of spacetime itself. In the past, the universe was smaller, denser, and more organized. In the future, it will be larger, more diffuse, and less organized. This cosmic evolution provides a universal timeline that applies throughout the observable universe.

Remarkably, the cosmological arrow appears to be aligned with the thermodynamic arrow. The expansion of the universe drives cosmic evolution toward states of higher entropy, from the highly organized conditions shortly after the Big Bang toward the cold, dilute future state when stars have burned out and black holes have evaporated. This alignment suggests a deep connection between local thermodynamic processes and global cosmic evolution.

The discovery of cosmological time dilation, which we explored in Chapter 12, provides direct observational evidence for this cosmological arrow. When we observe distant quasars appearing to evolve in slow motion, we're seeing the effects of cosmic expansion on the flow of time itself. The direction of increasing redshift—toward the future—coincides with the direction of increasing cosmic disorder and decreasing cosmic organization.

The Paradox of Microscopic Reversibility

The most puzzling aspect of time's arrow lies in what physicists call the paradox of microscopic reversibility. The fundamental laws governing individual particles are essentially time-symmetric—they work equally well whether time runs forward or backward. Yet these symmetric microscopic laws somehow give rise to irreversible macroscopic behavior.

This paradox becomes particularly acute when we consider that every macroscopic process—including the cooling of coffee, the aging of organisms, and the evolution of stars—is ultimately composed of microscopic interactions that are individually reversible. How can reversible microscopic processes combine to create irreversible macroscopic behavior?

The resolution lies in the statistical mechanics that underlies thermodynamics. While individual molecular collisions are perfectly reversible, the probability of vast numbers of molecules spontaneously coordinating their motion to reverse a macroscopic process is so small as to be effectively zero. The second law of thermodynamics emerges not as a fundamental law of nature but as a statistical consequence of dealing with astronomical numbers of particles.

Consider shuffling a deck of cards. Each individual card movement is perfectly reversible—you could, in principle, execute the exact reverse of every motion to return the deck to its original order. But the probability of randomly shuffling an already-shuffled deck back into its original sequence is so vanishingly small that it effectively never occurs. The thermodynamic arrow operates according to similar principles, but with numbers so large that the improbabilities become even more extreme.

Quantum Mechanics and Time's Direction

The relationship between quantum mechanics and time's arrow adds another layer of complexity to our understanding of temporal direction. The fundamental equation of quantum mechanics—the Schrödinger equation—is time-symmetric and doesn't inherently prefer one temporal direction over another. Yet quantum measurements appear to introduce irreversibility into quantum systems.

When a quantum system is measured, its wave function appears to "collapse" from a superposition of possible states into a definite outcome. This measurement process appears to be irreversible—we never observe quantum systems spontaneously "uncollapsing" from definite states back into quantum superpositions. This quantum mechanical arrow of time may be related to the thermodynamic arrow, but the connection remains a subject of active research.

Some interpretations of quantum mechanics suggest that the apparent irreversibility of measurement is actually a consequence of decoherence—the process by which quantum systems become entangled with their environments in ways that make quantum superpositions extremely difficult to maintain. From this perspective, the quantum arrow of time emerges from the same statistical principles that govern the thermodynamic arrow.

Entropy and Information

Modern developments in physics have revealed deep connections between entropy, information, and the arrow of time. The concept of entropy, originally developed in thermodynamics to describe the distribution of thermal energy, has been generalized to apply to any system where information can be organized or disorganized.

In this information-theoretic view, the arrow of time corresponds to the direction in which information becomes more difficult to access or recover. A scrambled egg contains the same information as an unscrambled egg, but that information has been distributed in a way that makes it practically impossible to recover. The process of scrambling represents an increase in entropy not because energy has been lost, but because information has been made inaccessible.

This connection between entropy and information has profound implications for understanding time's arrow in the context of Einstein's universe. The time dilation effects we've explored throughout this book don't change the fundamental direction of information flow—they only affect the rate at which information becomes disorganized relative to different observers.

Black holes provide a dramatic example of this connection. The extreme time dilation near a black hole's event horizon means that information falling toward the black hole appears to be frozen at the horizon from the perspective of outside observers. Yet the information continues to evolve toward higher entropy states according to observers following it inward, maintaining the consistency of time's arrow even under the most extreme gravitational conditions.

The Psychological Arrow

Our subjective experience of time's passage provides another manifestation of time's arrow that connects intimately to the physical arrows we've discussed. Human consciousness appears to have an inherent direction—we remember the past but not the future, we can influence future events but not past ones, and we experience a continuous flow of awareness from moment to moment.

This psychological arrow of time appears to be aligned with the thermodynamic arrow. Our brains are complex thermodynamic systems that operate by converting organized chemical energy into heat while processing information. The formation of memories represents a decrease in local entropy (information becomes more organized), but this decrease is more than compensated by increases in entropy elsewhere in the brain and body.

The time dilation effects described by Einstein's relativity don't appear to affect the psychological arrow of time. Astronauts experiencing time dilation still perceive time flowing in the normal direction—they still form memories of the past and anticipate the future. Their subjective experience of temporal flow remains aligned with the thermodynamic processes occurring in their brains, even though these processes appear slowed down to outside observers.

Biological and Evolutionary Arrows

Life itself provides a powerful example of how time's arrow manifests in complex systems. Biological evolution represents a process that appears to create order and complexity over time, seemingly opposing the thermodynamic tendency toward disorder. Yet evolutionary processes actually operate within the framework of the second law of thermodynamics.

Organisms maintain their complex organization by consuming energy from their environment and converting it into heat, thereby increasing the total entropy of the universe while decreasing their local entropy. Evolution proceeds by selecting for organisms that are most efficient at this entropy-increasing process, creating the appearance of increasing complexity while actually accelerating the universe's overall tendency toward disorder.

The biological arrow of time is intimately connected to the thermodynamic arrow. Aging, growth, metabolism, and reproduction all represent irreversible processes that proceed in the direction of increasing entropy. Even the most sophisticated biological systems—including human brains—operate as entropy-increasing engines that create local order at the expense of global disorder.

Implications for the Future

Understanding time's arrow has profound implications for humanity's future as we expand into Einstein's relativistic universe. As we develop technologies that exploit extreme time dilation effects—whether through high-speed travel or strong gravitational fields—we must remember that these technologies operate within the framework of thermodynamic irreversibility.

Relativistic spacecraft will still experience aging and entropy increase, though these processes may appear slowed down to external observers. The electronics, life support systems, and even the crew members aboard such spacecraft will continue to evolve toward higher entropy states, maintaining the consistency of time's arrow even under extreme relativistic conditions.

The exploration of black holes and other extreme gravitational environments will provide new tests of how time's arrow behaves under conditions of extreme time dilation. Near the event horizons of black holes, where time dilation becomes infinite from external perspectives, the relationship between thermodynamic and gravitational effects may reveal new aspects of temporal direction that we don't yet understand.

The Deep Unity of Physics

The arrow of time reveals a deep unity underlying the diverse phenomena we've explored throughout this book. From the GPS satellites that require relativistic corrections to maintain accuracy, to the extreme time dilation near black holes, to the cosmic time dilation observed in distant quasars, all of these effects occur within a universe where time has a fundamental direction determined by the increase of entropy.

This unity suggests that time's arrow isn't just a curious feature of thermodynamics but a fundamental aspect of how the universe works. The same statistical principles that explain why coffee cools also explain why the universe expands, why stars evolve, and why life exists. The direction of time emerges from the deepest levels of physical reality and manifests in every process we observe.

Einstein's insights about the malleability of time don't eliminate time's arrow—they reveal how temporal direction interacts with temporal distortion in ways that preserve the fundamental asymmetry of temporal flow. Whether time passes quickly or slowly, whether it's stretched by gravity or compressed by motion, it continues to flow in the direction of increasing entropy, maintaining the arrow that gives meaning to concepts like past, present, and future.

The Continuing Mystery

As Dr. Foster contemplated his long-cold coffee, he reflected on how the simple act of cooling embodied one of the deepest mysteries in physics. The arrow of time that prevented his coffee from spontaneously reheating was the same arrow that drives cosmic evolution, biological development, and the formation of memories in conscious minds.

Yet despite decades of research, fundamental questions about time's arrow remain unanswered. Why does the universe appear to have started in a highly organized state that could evolve toward disorder? Is time's direction truly fundamental, or does it emerge from more basic physical principles? How does the arrow of time relate to the quantum mechanical aspects of reality that we explored in our previous chapter?

These questions continue to drive research at the frontiers of physics, connecting studies of black holes and cosmology to investigations of consciousness and the nature of information. The arrow of time serves as a bridge between the abstract mathematics of Einstein's field equations and the everyday experience of living in a universe where past and future have meaning.

The story of time's arrow reminds us that even as we learn to navigate Einstein's universe of curved spacetime and relativistic effects, we remain embedded in a cosmos where time has direction and meaning. The clockmaker's dilemma that opened our journey has been replaced by a deeper understanding of time's nature, but that understanding reveals new mysteries about why time flows as it does and what determines its relentless march toward the future.

In discovering that time has an arrow, we've learned that the universe itself has a kind of memory—a preferred direction that connects the organized past to the disordered future. This arrow provides the backdrop against which all the time dilation effects we've explored take place, ensuring that even as time bends and stretches under the influence of gravity and motion, it continues to point steadfastly toward tomorrow.

The arrow of time stands as a reminder that physics isn't just about mathematical equations and experimental measurements—it's about understanding the deep principles that make existence possible and meaningful. In a universe without time's arrow, there would be no evolution, no life, no consciousness, and no science. The thermodynamic tendency toward disorder that prevents coffee from spontaneously reheating is the same tendency that enabled stars to form, planets to develop, and minds to emerge that could discover the arrow of time itself.

As we continue to explore the implications of Einstein's insights about space and time, we carry with us the profound realization that time's arrow provides the temporal framework within which all physical processes unfold. From the quantum scale to the cosmic scale, from the briefest particle interactions to the longest astronomical cycles, everything occurs within the flow of time that moves inexorably from past to future, driven by the fundamental tendency of the universe to evolve from order toward disorder.

This arrow ensures that our journey through Einstein's universe is not just a tour through abstract physical principles but a voyage through a cosmos where time has meaning, where causes precede effects, and where the future remains genuinely open to possibilities that have not yet been realized. In understanding time's arrow, we come to appreciate not just how time can bend and stretch, but why it flows at all—and why that flow makes the universe a place where stories can be told, discoveries can be made, and the arrow of time itself can be comprehended by minds that emerged from the very processes it describes.

Chapter 18: Living in Einstein's Universe

As our journey through the twisted landscapes of time and space draws to a close, we find ourselves transformed. We began as inhabitants of Newton's clockwork universe, where time ticked steadily forward like a cosmic metronome, indifferent to motion or gravity. We end as citizens of Einstein's universe, where time itself dances to the rhythm of velocity and bends under the weight of mass and energy.

This transformation is not merely intellectual. It represents a fundamental shift in how we understand our place in the cosmos and our relationship with the very fabric of reality. We no longer live in a universe of absolute time and space, but in one where these most basic concepts are relative, dynamic, and intimately connected to the dance of matter and energy around us.

The Invisible Architecture of Modern Life

Every moment of our daily existence unfolds within Einstein's relativistic framework, even when we remain blissfully unaware of it. The smartphone in your pocket depends on GPS satellites that must account for both special and general relativistic effects to maintain their precision. Without Einstein's corrections, these satellites would drift by kilometers each day, rendering navigation impossible.

The electricity powering your home flows through a grid that operates on principles of electromagnetic induction, phenomena that are fundamentally relativistic in nature. The very reason copper wires conduct electricity while plastic insulates involves the relativistic behavior of electrons in different materials. Even the golden color of wedding rings emerges from relativistic effects that contract the electron orbitals in gold atoms.

In hospitals around the world, PET scans reveal the inner workings of living brains by detecting the annihilation of electrons and positrons, matter and antimatter particles whose existence was predicted by the relativistic marriage of quantum mechanics with Einstein's special relativity. The linear accelerators used in cancer treatment accelerate particles to near light speed, where relativistic effects become crucial for precise targeting of tumors.

Time as Our Companion

Perhaps most remarkably, we have learned that time itself is not the impartial observer we once believed it to be. Time is deeply personal, intimately connected to our motion through space and our position in gravitational fields. The time you experience as you read these words is uniquely yours, slightly different from that of someone at a different altitude, moving at a different velocity, or simply in a different gravitational environment.

This revelation transforms our relationship with time from passive observation to active participation. We are not merely carried along by time's current; we shape and mold our personal timestreams through our choices and movements. When you drive to work, you experience time fractionally slower than someone at rest due to your motion. When you climb a mountain, you age infinitesimally faster than someone at sea level due to the weaker gravitational field. These effects are tiny in our everyday experience, but they are real and measurable.

The precision of atomic clocks has made these philosophical concepts concrete and practical. Time is no longer abstract but has become a technology, a resource that can be measured, distributed, and synchronized across the globe with extraordinary precision. The international atomic time standard represents humanity's collective effort to create a universal temporal reference, even while recognizing that each observer's experience of time remains fundamentally unique.

The Cosmic Perspective

Living in Einstein's universe means embracing a cosmic perspective that extends far beyond our earthly experience. We have seen how time dilation creates windows into cosmic history, allowing us to observe distant quasars and galaxies as they existed billions of years ago. The light from these ancient objects carries information about time's passage in environments radically different from our own.

Near the supermassive black hole at the center of our galaxy, time crawls compared to our Earth-bound perspective. Observations of stars orbiting close to this cosmic monster provide direct evidence of gravitational time dilation in one of the most extreme environments in the universe. These observations connect our local understanding of time with the grand cosmic theater in which we play our small but significant part.

The cosmic microwave background radiation that fills the universe provides another temporal reference point. This afterglow of the Big Bang reaches us from a time when the universe was only 380,000 years old, carrying information about cosmic conditions when time itself was young. In studying this radiation, we probe not just the history of space and matter, but the evolution of time itself as the universe expanded and cooled.

The Philosophy of Relativistic Existence

Einstein's discoveries force us to confront fundamental questions about the nature of reality and our place within it. If time and space are relative, what aspects of existence are truly absolute? Einstein himself identified the speed of light in vacuum as one such absolute, a universal constant that provides the scaffolding upon which all relativistic phenomena hang.

But perhaps the deeper constant is the mathematical structure underlying these relationships. The Lorentz factor $\gamma=1/\sqrt{(1 - v^2/c^2)}$ that governs time dilation is more than just an equation; it represents a fundamental symmetry of nature. This mathematical harmony suggests that the universe operates according to profound geometric principles that transcend our everyday intuitions about space and time.

The equivalence principle, which equates gravitational and inertial mass, reveals another deep truth about the universe's construction. This principle suggests that gravity is not a force in the traditional sense but rather a manifestation of spacetime's curvature. We live in a universe where the stage on which physics plays out is itself a dynamic participant in the drama.

Quantum Gravity and the Future of Understanding

As we have explored, Einstein's classical general relativity faces challenges at the smallest scales, where quantum mechanics becomes important. The search for a theory of quantum gravity represents one of the greatest intellectual adventures of our time. The resolution of this challenge will likely transform our understanding of time, space, and reality as profoundly as Einstein's original insights did over a century ago.

String theory suggests that the familiar three dimensions of space and one dimension of time might be merely the visible surface of a reality with many more dimensions. Loop quantum gravity proposes that space and time themselves are granular at the smallest scales, made of discrete units too small for us to perceive directly. These theories push us toward even stranger concepts: perhaps time itself emerges from more fundamental timeless processes, or perhaps the arrow of time is not a universal feature but depends on the perspective of the observer.

The experimental verification of gravitational waves has opened new windows into extreme relativistic phenomena. These ripples in spacetime itself, detected by facilities like LIGO, allow us to observe the universe through a completely new sense. We can now "hear" black holes colliding billions of light-years away, experiencing the actual stretching and compression of space and time as these cosmic catastrophes unfold.

The Educational Revolution

Einstein's universe has transformed not just physics but education itself. The concepts we have explored in this book were once accessible only to graduate students and professional physicists. Today, precision atomic clocks and GPS satellites have made relativistic effects measurable in high school laboratories. Students can observe time dilation directly using fast-moving particles in cloud chambers, or calculate the relativistic corrections needed for accurate navigation.

This democratization of relativistic physics represents a broader cultural shift. We live in an era where the most exotic predictions of twentieth-century physics have become twenty-first-century technology. The weird and wonderful phenomena that once lived only in the imaginations of theoretical physicists now power the devices we carry in our pockets and the satellites that orbit overhead.

The educational challenge is no longer convincing students that these effects exist, but helping them understand why they were so difficult to discover and accept. The history of relativity teaches us about the conservative nature of human intuition and the courage required to abandon comfortable assumptions about reality when evidence demands it.

Technology and the Relativistic Future

As we look toward the future, relativistic effects will become increasingly important in human technology and exploration. Plans for interstellar missions must account for time dilation effects that could separate space travelers from Earth not just by distance but by time itself. A journey to nearby stars at relativistic speeds would return explorers to an Earth decades or centuries older than when they departed.

The development of fusion energy depends on our understanding of the mass-energy equivalence $E = mc^2$ that transforms tiny amounts of matter into enormous quantities of energy. Quantum computers rely on principles that emerge from the marriage of quantum mechanics with special relativity. Even artificial intelligence systems designed to navigate in real-time must account for the finite speed of light and relativistic signal delays.

The Global Positioning System represents just the beginning of humanity's relativistic technological infrastructure.

Future space-based civilizations will require even more sophisticated temporal coordination, accounting for the complex gravitational fields of planets, moons, and artificial habitats. Time itself will become a carefully managed resource, distributed and synchronized across the solar system with precision that makes today's atomic clocks seem primitive.

The Environmental Dimension

Einstein's universe also provides new perspectives on environmental and sustainability challenges. The mass-energy equivalence that powers the sun also reveals the enormous energy density available through fusion reactions. Understanding stellar evolution through relativistic physics shows us both the ultimate limits of our sun's lifetime and the potential for harnessing similar processes here on Earth.

The precise measurements enabled by relativistic physics allow us to track climate change with unprecedented accuracy. Atomic clocks can detect sea level rise by measuring gravitational time dilation changes as Earth's mass distribution shifts. GPS satellites monitor ice sheet thickness, ocean currents, and atmospheric dynamics with precision that would be impossible without relativistic corrections.

Gravitational wave astronomy opens new possibilities for understanding Earth's place in the cosmic environment. These detectors can sense disturbances in spacetime itself, potentially providing early warning of cosmic catastrophes or revealing previously unknown sources of gravitational waves throughout the universe.

The Human Element

For all its mathematical sophistication and technological applications, Einstein's universe remains fundamentally human in its implications. The relativity of simultaneity means that there is no universal "now" shared across the cosmos, yet this same principle allows for the possibility of genuine novelty and creativity in the universe. The future is not predetermined in some cosmic reference frame, but emerges through the complex interactions of matter, energy, and spacetime.

The personal nature of time means that each human life represents a unique trajectory through spacetime, with its own proper time and particular sequence of events. Your biological aging follows your personal worldline through the curved spacetime around Earth, slightly different from every other person's temporal journey. In this sense, relativity makes each life cosmically unique.

The arrow of time that governs our psychological experience aligns with the thermodynamic arrow that drives the evolution of stars, galaxies, and the universe itself. We are not separate from the cosmic processes we observe but integral participants in the universe's ongoing exploration of its own possibilities.

The Continuing Mystery

As our journey ends, we must acknowledge that many profound questions remain open. Why does the universe have the particular relativistic structure that Einstein discovered? Why do the fundamental constants have the values they do? What determines the arrow of time, and could it flow in the opposite direction under different circumstances?

The relationship between consciousness and time remains particularly mysterious. How does the subjective experience of temporal flow relate to the objective geometric structure of spacetime? Why do we remember the past but not the future, when the fundamental laws of physics are largely time-reversible? These questions sit at the intersection of physics, neuroscience, and philosophy, awaiting future generations of investigators.

Even more fundamentally, we do not yet understand why mathematics proves so unreasonably effective at describing the structure of spacetime. The mathematical beauty of Einstein's field equations seems to capture something essential about the universe's deepest nature, but the reason for this correspondence remains mysterious.

A Universe of Wonder

Living in Einstein's universe means inhabiting a reality far stranger and more wonderful than our ancestors could have imagined. We exist in a cosmos where space and time are woven together into a dynamic fabric that can stretch, twist, and ripple. We ride a planet that follows curved geodesics through this spacetime, creating the phenomenon we experience as gravity.

Every second of our lives, we age at a rate determined by our motion through space and our position in gravitational fields. The atoms in our bodies were forged in the nuclear furnaces of ancient stars, through processes governed by the same mass-energy equivalence that makes atomic clocks tick and GPS satellites navigate. The light we see from distant galaxies left them billions of years ago, carrying information about cosmic conditions when the universe was young and time itself was different.

This perspective should inspire both humility and wonder. We are small creatures on a small planet orbiting an ordinary star in an unremarkable galaxy among billions of others. Yet we are also conscious beings capable of understanding the fundamental structure of spacetime itself, able to measure time's passage with extraordinary precision and probe the universe's most exotic phenomena.

The Eternal Journey

The title of our book speaks of time bending, but perhaps the most remarkable thing we have learned is that this bending is not a deviation from normalcy but the very foundation of reality. Time does not bend because something is wrong with it; time bends because this flexibility is its essential nature. Space and time are not rigid containers within which events unfold, but dynamic participants in every physical process.

As we close this exploration, remember that you continue to live in Einstein's universe every moment of every day. The time you experience as you put down this book is your own, shaped by your motion and gravitational environment. The light reaching your eyes left its sources at various times in the past, creating a complex temporal collage of information from different epochs.

The universe continues to expand, carrying distant galaxies away from us and stretching the wavelength of their light toward the red end of the spectrum. Gravitational waves from cosmic collisions continue to ripple through spacetime, stretching and compressing the space between atoms in detectors around the world. The cosmic microwave background continues to cool as the universe ages, providing an ever-changing record of cosmic history.

In Einstein's universe, we are all time travelers, moving inexorably forward through the fourth dimension at the rate of one second per second of proper time. We are all participants in the grand cosmic experiment of spacetime, contributing our own small but unique worldlines to the universe's ongoing story. The elastic nature of time ensures that this story will continue to surprise and delight future generations of explorers, as they discover new ways to bend, stretch, and navigate the magnificent temporal landscape that Einstein revealed to us over a century ago.

The journey through curved spacetime continues, and we are privileged to be conscious passengers on this remarkable ride through the cosmos, where time truly bends and reality proves far stranger than fiction.

www.ingramcontent.com/pod-product-compliance
Lightning Source LLC
LaVergne TN
LVHW052309200926
848481LV00002B/511

* 9 7 9 8 3 4 9 5 7 4 3 6 8 *